# Bioseparation Process in Foods

# Bioseparation Process in Foods

*Editor*

*P.B. Bansal*

2012

**Gene-Tech Books**

**New Delhi - 110 002**

**ISBN 81-89729-03-9**

*Published by* : **GENE-TECH BOOKS**
**4762-63/23, Ansari Road, Darya Ganj,**
**NEW DELHI - 110 002**
**Phone: 41562849**
**e-mail: genetechbooks@yahoo.co.in**

*Printed at* : **Tarun Offset Printers**
**Delhi**

**PRINTED IN INDIA**

# Preface

The bioseparation techniques are increasingly important for biotechnology because separation is often the limiting factor for the success of biological processes. Manufacturers of new enzymes and pharmaceutical products require improved methods for recovering intact cells and intracellular products. Similarly the isolation, purification and concentration of many biomolecules produced in fermentation processes is extremely important. Often such downstream processing contributes a large portion of the product cost and thus efficient and economical alternative approaches to bioseparation processes are needed to eliminate, reduce or facilitate the handling of solids. Bioprocessing is an important reference work and is a comprehensive description of the state-of-the-art of bioprocessing, covering traditional and newly introduced processes.

This book describes the basic principles and engineering design of various separation processes including chromatography, electrophoresis, extraction, crystallization, and membrane separation. It will stimulate young engineers and scientists who will develop bioseparation engineering further in the 21st century, and contribute to a world-wide attention to the global environment.

The book, structured in 275 pages, was elaborated by the co-operation of many contributors coming from several countries and aimed to bring the latest information in the field of bioseparation of enzymes and proteins. It aims to systematically promote and enforce an essential integration between bioseparation processes and biological/chemical and engineering sciences through a series of case studies focusing on separations and separation processes in the context of industrial practice.

***Editor***

# Contents

# 1

# Bioseparation Process

The rapid growth in the field of modern biotechnology along with fast commercialization of biotechnological products has led to an increase in the demand for efficient, large-scale purification processes for biologically derived products. Bioseparations engineering is challenging primarily due to the following reasons:

1. Biotechnological processes usually yield products at very low concentrations in the product stream.
2. The product of interest needs to be separated from a large number of impurities, some of which have physical and chemical properties not too different from the product.
3. Bioproducts tend to be thermolabile. Therefore thermal separation processes cannot be used.
4. Most bioproducts are stable only within particular pH and ionic strength ranges. This limits the operating conditions that may be used in the separation process.
5. Certain bioproducts are shear sensitive.
6. A large number of bioproducts cannot be dissolved in organic solvents or are unstable in their presence.

7. Most bioproducts have very stringent quality requirements both in terms of percentage purity and absence of specific impurities.

These factors imply that an ideal bioseparation process must combine high throughput along with high selectivity of separation, and should be carried out under "mild" operating conditions.

### Biological Products

A large number of products are manufactured by biological processes and hence need to be "bioseparated". Table 1 lists some of the common biologically derived products.

*Table 1. Biologically derived products*

| Product | Nature of bioseparation required |
|---|---|
| *Alcoholic beverages*: | |
| Beer, wine, spirits | Clarification, distillation |
| *Organic acids*: | |
| Acetic acid, citric acid | Precipitation, filtration, adsorption, solvent extraction |
| *Vitamins*: | |
| Vitamin C, vitamin B12, riboflavin | Precipitation, filtration, adsorption, solvent extraction |
| *Amino acids*: | |
| Lysine, glycine, phenyl alanine | Precipitation, filtration, adsorption, solvent extraction |
| *Antibiotics*: | |
| Penicillins, neomycin, bacitracin | Precipitation, filtration, adsorption, solvent extraction |
| *Carbohydrates*: | |
| Starch, sugars, dextrans | Precipitation, filtration, adsorption |
| *Lipids*: | |
| Glycerol, fats, fatty acids | Precipitation, filtration, adsorption, solvent extraction |

| | |
|---|---|
| *Proteins:* | |
| Food and food additives<br>Nutraceuticals<br>Industrial enzymes<br>Hormones<br>Pharmaceutical enzymes<br>Plasma derived products<br>Monoclonal antibodies<br>Growth factors<br>Clotting factors<br>Thrombolytics<br>r-DNA derived proteins<br>Diagnostic proteins<br>Vaccines | Filtration, precipitation, centrifugation, adsorption, chromatography, membrane based separations |
| *DNA based products:* | |
| DNA probes, plasmids, nucleotides, oligonucleotides | Filtration, precipitation, centrifugation, adsorption, chromatography, membrane based separations |

## Requirements for Bioseparation

— Purity

— Stability

— Cost

— Reproducibility

— Scalability

### *Economic Importance*

The isolation of biotechnological products from bioreactor medium and other biotechnological feed streams has long been recognized as a technical difficulty and expensive procedure. Downstream processing is often the critical limiting factor in the development of commercial biotechnology. Isolation and purification cost is a substantial fraction of the total cost of production. Table 2 shows the downstream processing cost as approximate proportion of the selling price of certain biotechnological products.

*Table 2. Cost of bioseparation*

| Product | Approximate relative price | Downstream processing cost (%) |
|---|---|---|
| Ethanol | 1 | 15 |
| SCP | 0.8 | 20 |
| Yeast biomass | 2 | 20 |
| Citric acid | 3.2 | 30-40 |
| Monosodium glutamate | 5 | 30-40 |
| Xanthan | 20 | 50 |
| Penicillin G | 60 | 20-30 |
| Bulk enzymes | 100 | 40-65 |
| Therapeutic proteins/DNA | >500 | 60-80 |

For products such as therapeutic proteins and nucleic acids the downstream processing cost is the major cost and hence most of the innovations in bioseparations are taking place in the area of protein and nucleic acid purification. Quite clearly, the manufacturing cost for these products could be significantly reduced by cutting down the bioseparation cost.

### *Strategies for Bioseparation*

A large number of purification techniques are available and when purifying a specific product, a purification procedure has to be developed using these techniques in particular combinations. Process design and development should take into consideration the following:

1. The volume of process stream
2. The relative abundance of the product in this process stream

3. The intended use of the product, i.e. purity requirements
4. The cost of the product
5. Stability requirements

A RIPP (Recovery, Isolation, Purification and Polishing) scheme is commonly employed for bioseparation. This strategy involves using low-resolution techniques (e.g. precipitation, filtration, centrifugation, crystallization) first followed by high-resolution techniques (e.g. adsorption, chromatography, electrophoresis). Biotechnological processes usually yield target products at very low concentrations in the product stream. The product of interest also needs to be separated from a large number of impurities, some of which have physical and chemical properties not too different from the product.

The conventional way of thinking ran this way: if an impure product stream were to be sent directly to a high-resolution separation device, it would soon be overwhelmed and hence fail to function properly. Therefore it made sense to use high throughput-low resolution techniques first to significantly reduce the volume/concentration of process stream and enrich it with respect to the target product. This high throughput-low resolution part of a conventional separation process is referred to as recovery and isolation (i.e. RI).

The partially processed product stream would then be further processed by high resolution-low throughput techniques to get the pure finished target products, this part being referred to as purification and polishing (i.e. PP). With the advent of membrane separation processes it is now possible to replace the conventional RIPP scheme. Membrane processes give high throughput and can be fine-tuned/optimized to give very high resolution. The use of membrane technology reduces the number of bioseparation steps and hence contributes towards high product recovery.

Three lists of bioseparation methods are given below:

— Low resolution-high throughput

1. Cell disruption
2. Precipitation
3. Centrifugation
4. Liquid-liquid extraction
5. Leaching
6. Filtration
7. Supercritical fluid extraction
8. Microfiltration
9. Dialysis

— High resolution-low throughput

1. Ultracentrifugation
2. Adsorption
3. Packed bed chromatography
4. Affinity separation
5. Electrophoresis

— High resolution-high throughput

1. Ultrafiltration
2. Fluidized bed chromatography
3. Membrane chromatography

### Processing of Fruit and Vegetable

The fruit and vegetable processing industries may be

described as consisting of two segments: fresh pack and processing. The former collects crops and field packs them into lug boxes or bulk bins for shipment to a produce finishing plant. Crops are cooled to preserve integrity and fumigated or treated to control insect infestation or microbial disease development. The processing segment, or packers, includes all unit operations, extending the shelf life of food being processed and adding value through produce modification to satisfy market niches.

The fresh pack segment of the industry shares unit operations with the processing segment. These operations are the sorting/trimming, washing, grading, and packing lines. But after the packing lines, additional unit operations may add to the waste generating scheme for the processing segment alone. Additional operations may include combinations of peeling, stemming, snipping, pitting, trimming, chopping, and blanching. In some instances, the final product is dehydrated (e.g., chopped onions). In others, it is packaged and processed.

Processing can include one treatment or a combination of several treatments (e.g., acidifying, brining, freezing, or cooking). Major water use and waste generation points associated with the fruit and vegetable industry include the washing steps for raw and processed produce, peeling and pitting practices, blanching, fluming the produce after blanching, sorting, and conveying the product within the plant. Reducing size, coring, slicing, dicing, pureeing, and juicing process steps, as well as filling and sanitising activities after processing, also contribute to the wastestream.

Major wastewater characteristics to be considered for the vegetable and fruit processing industry are the wide ranges of wastewater volume and the concentrations of organic materials. Wastewater characteristics can be influenced by a number of factors such as the commodity processed, the process unit operations used, the daily-production performance level, and the seasonal variation, e.g., growing condition and crop age at harvest. In the processing environment for vegetable and fruit material handling, heating, cooling, and packaging, there are six major contributing point sources for waste. These sources are the following operations:

(1) raw produce washing, grading, and trimming,

(2) washing after steam/lye peeling and/or size reducing,

(3) blanching and fluming,

(4) filling,

(5) sanitation/ plant cleanup, and

(6) processed product cooling.

Plant management practices greatly influence process operation efficiency relative to final product yield and waste quantity generated. Ideally, considerable waste reduction can be achieved if harvesting equipment permits additional stems, leaves, and culled materials to remain in the field during harvest. If crop washing, grading, and trimming can occur in the field, then additional soil and food residues will remain at the farm. Realistically, most such wastes are being handled at vegetable and fruit processing plant sites. Primary waste-management strategies used by this industry are water conservation and waste-solids separation. Water use by the vegetable and fruit processing industry is essential to the washing, heating, and cooling of food products. But the industry has adopted a number of practices, showing heightened sensitivity to the need for water conservation:

1. Use of air flotation units to remove suspended debris from raw crop materials

2. Recovery and reuse of process water throughout the processing plant.

3. Decrease of water volume use in peeling and pitting operations, as well as decrease of raw product losses.
4. Separation of waste process streams at their sources, for potential by-product use.
5. Countercurrent reuse of wash and flume/ cooling waters.
6. Separation of low and high strength wastestreams.
7. Installation of low-volume, high-pressure cleanup systems.
8. Conversion from water to steam blan ' ing.
9. Use of air cooling after blanching.

The initial preparation processes for canned, frozen, and fermented fruits are washing, sorting, trimming, peeling, pitting, cutting or slicing, inspecting, and grading. Unwanted and undesirable materials must be removed before the fruits undergo additional processing, but not all fruits are subject to each step. For example, cherries and plums may be canned whole and unpeeled whereas apples, peaches, and pears must be peeled and either cored or pitted before being canned. Peeling can be by hand or with machines, chemicals, or steam.

After inspection and grading, the peeled fruits are conveyed mechanically or flumed to product handling equipment for processing. The converted fruit handling processes are can filling, syrup adding, exhausting and sealing, thermoprocessing, can cooling, and storing. Processing equipment and plant floor usually are cleaned at the end of each shift and so constitute a final source of waste materials.

Several water conservation and waste prevention techniques are available by which to decrease water volume. These techniques include

- The use of high-pressure sprays for clean-up.
- The elimination of excessive overflow from washing and soaking tanks.
- The substitution of mechanical conveyors for flumes, the use of automatic shut-off valves on water hoses.
- The separation of can cooling water from composite wasteflow.
- The recirculation of can cooling water. When can cooling water is not recirculated, it may be reused in caustic soda (NaOH) or in water peeling baths, in removal of NaOH after peeling, in primary wash of the raw material, in canning belt lubrication, and in plant cleanup operations.

### Processing of Dairy Products

The processing of dairy products often entails various unit operations. These generally include the receiving and the storing of raw materials, the processing of raw materials into finished products, the packing and the storing of finished goods, and a number of ancillary processes (e.g., heat transferring and cleaning) associated indirectly with processing and distributing. Equipment and facilities for receiving, transporting, and storing raw materials are much the same industrywide. Bulk carriers unload products in receiving areas by means of flexible lines or dump material into hoppers connected to fixed lines subsequently transferred by pump to storage.

Storage facilities can be of the refrigerator, vertical, or silo type, with storage tanks containing either liquid or dry products and ranging in volume from a few thousand gallons to one million gallons or more. Milk, a perishable product made up of fat, protein, carbohydrates, salts, and vitamins, is an ideal food for microorganisms as well as for humans. Thus, it needs to be protected from contamination, and much of the efforts of the dairy industry are directed to this end. Milk and its by-products are processed according to approved procedures, on machinery normally run no longer than about 20 hours per day.

Much equipment is dismantled daily. Systems may be cleaned in place or after they are taken apart. Automated cleaning systems, now predominant in the industry, require less labor but more water and cleaning chemicals than hand washing dismantled equipment does. Dairy processing wastewaters are generated during the pasteurisation and the homogenisation of fluid milk and the production of dairy products such as butter, ice cream, and cheese. The principal constituents of these wastewaters are whole and processed milk, whey from cheese production, and cleaning compounds. Water use in the dairy products industry depends on plant complexity and water-management practices.

Process wasteloads also differ considerably and are influenced greatly by the extent to which the plant controls raw material and product losses. Milk product losses typically range from 0.5 percent in large, technologically advanced plants to greater than 2.5 percent in small, old plants. Given redoubled effort by management, water usage in most plants could be decreased to approximately 0.50 L/kg milk equivalent processed. Considerable improvements in water and waste management remain important and realistic industry goals.

In recent years, technological innovations with membrane systems have provided many new opportunities. For example, ultrafiltration now can be used instead of the biological separation of organic material from liquid substrate. And instead of using reverse-osmosis systems for tertiary waste treatment, some food plants use them to recycle internal liquid wastestreams. The outflow from reverse-osmosis treatment can be of better quality than the native water.

### Processing of Meat and Poultry

The meat and poultry processing industries in the United States together make up a $75.6 billion per year industry. The U.S. Department of Commerce reported that the value of red meat shipments for 1988 totaled $46.8 billion. Most red meat processing plants are located in the Midwest; most poultry processing plants are in the Southeast and the Mid-Atlantic. Processing of prepared meats, including canned cooked products, luncheon meats, hot dogs, bacons, stews, and other ready-to-eat meat products, has expanded rapidly in recent years.

Most waste products are recovered somehow by the industry. Blood, feathers, and bone usually are processed into a meal product for animal feed. Similarly, meat scraps unsuitable for processing into food products are sold or given to rendering facilities for processing into animal and pet foods. The ultimate characteristics of solid materials and wastewaters generated by these source areas in a plant and unrecovered for another use differ greatly and are affected by:

1. animal size and type
2. processing level
3. conveyance means
4. processing water use
5. cleanup and housekeeping procedures

Water use for broiler processing typically ranges from 3.5 to 10.0 gal./bird; for turkeys, 11 to 23 gal./bird. Flow rates of 350 gal./animal have been reported for beef slaughtering plants. In one beef slaughtering operation, water use dropped from 458 to 187 gal/ head after water conservation measures were adopted. Water is used for chilling, scalding, can retorting, washing, cleaning, and waste conveying. For example, poultry processing uses approximately 3.5 to 7.0 gallons of water per bird of four-pound average weight. All broiler processing plants are required to have a scalder overflow rate of 0.25 gal./bird and a chiller overflow rate of 0.50 gal./bird.

In many instances, this water is used in the plant for the transport of feathers and offal from the processing area. One researcher, studying a broiler processing plant, reported that processing accounted for 76 percent of the water use, with 13 percent used in cleanup and 12 percent used in downtime. Beef processing water usage, primarily from carcass washing and process clean-up, has been reported in the range of 150 to 450 gallons per animal processed.

As a general rule, meat processors use about one gallon of water per pound of processed hamburger meat. The amount of wastewater generated by the industries can be decreased largely through changes in cleanup practices. Water use can be minimised by means of commercially available high-pressure, restricted flow hoses, which can be fit with automatic shutoffs to prevent water loss during inactivity. Many materials can be handled mechanically. For example, flour and other dry material can be vacuumed from the floor and augers and conveyors can be used to transport scrap meat and viscera.

Chiller and scalder water is reused in most poultry processing plants for flushing water to remove offal and feathers. Reconditioning of chiller overflow through the use of filtration and ultraviolet irradiation has been recommended. Limits to use include the potential of bacterial contamination by coliforms or by Escherichia coli. Recycling is limited by the characteristics of the wastestream and by the potential for contamination of food products.

### Oil Processing

The extracting, refining, and processing of edible oils produces a variety of waste products. Separate totals are presented with and without salad dressing and mayonnaise because these processes often are absent in a facility. Certain oil processing and refining operations have no oil seed processing facilities, but instead bring in crude vegetable oil. Obviously, operations of an atypical size or those omitting certain processes will have different waste loads. This applies especially to operations involved in acidulation or in mayonnaise and salad dressing processing.

These loadings are representative for an operation running reasonably well from a process loss control standpoint. But actual loadings depend on how well plants are run. A final source of wastewater is contaminated runoff from truck and rail loadout areas and from tank farm drainage. During rainy periods, runoff from these sources can contribute the equivalent of five to ten gal/min to total daily average flow and, in fact, may affect peak flows to a much greater extent.

### Food and Beverage

In the food and beverage industry, water plays a significant role in transporting, cleaning, processing, and formulating products as well as meeting many federal sanitary standards. Facilities implementing water conservation programmes sometimes struggle to balance these needs with the many benefits of reducing water usage. Several opportunities in the beverage industry include:

— Adjust pump cooling and flushing water to the minimum required.

— Investigate potentially reusable discharges including final rinses from tank cleaning, keg washes, fermenters; bottle and can soak and rinse water; cooler flushwater; filter backwash; and pasteriser and steriliser water.

— Potential areas for reuse include first rinses in wash cycles; can shredder; bottle crusher; filter backflush; caustic dilution; boiler make-up; refrigeration equipment defrost; and floor and gutter wash.

Several opportunities in the food industry include:

— Recycle transport water where feasible.

— Use conveyor belts for product transport. Preference should be given to ìrabbit-earî or V-shaped roller supports because these are much easier to clean.

— Use pneumatic conveying systems where practical.

— Use flumes with parabolic cross-sections rather than flat bottom troughs.

— Consider these alternatives to water-intensive units: 1) rubber-disc units instead of raw product cleaning and peeling; 2) steam instead of water blanchers, 3) evaporative coolers instead of water cooled systems.

— Establish optimum depth of product on conveyors to maximise wash water efficiency.

— Divide spray wash units into two or more sections, and establish a counterflow reuse system.

— Control belt sprays with a timer to allow for intermittent application of chlorinated water.

# 2

# Bioprocess Design

Process design is the conceptual work done prior to building, expanding or retrofitting a process plant. It consists of two main activities, process synthesis and process analysis. Process synthesis is the selection and arrangement of a set of unit operations (process steps) capable of producing the desired product at an acceptable cost and quality.

Process analysis is the evaluation and comparison of different process synthesis solutions. In general, a synthesis step is usually followed by an analysis step, and the results of analysis determine the subsequent synthesis step. Process design and project economic evaluation require integration of knowledge from many different scientific and engineering disciplines and are carried out at various levels of detail. Table 1 presents a common classification of design and cost estimates and typical engineering cost for a \$50 million plant.

Figure 1 presents the need for various types of design estimates during the lifecycle of product development and commercialisation. The trapezoidal shape of the graph represents the drastic reduction in product candidates as we move from feasibility studies to commercialisation. In fact, the chances of commercialisation at the research stage for a new product are only about 1 to 3%, at the development stage they are about 10 to 25%, and at the pilot plant stage they are about 40 to 60%.

*Table 1. Types of design estimates*

| Level | Type of Estimate | Accuracy | Cost ($1000) |
|---|---|---|---|
| 1. | Order-of-Magnitude estimate (ratio estimate) based on similar previous cost data. | ≤ 50% | |
| 2. | Project Planning estimate (budget estimation) based on knowledge of major equipment items. | ≤ 30% | 20 - 40 |
| 3. | Preliminary Engineering (scope estimate) based on sufficient data to permit the estimate to be budgeted. | ≤ 25% | 50 - 100 |
| 4. | Detailed Engineering (capital approval stage) based on almost complete process data. | ≤ 15% | 100 - 200 |
| 5. | Procurement and Construction (contractor's estimate) based on complete engineering drawings, specifications, and site surveys. | ≤ 10% | 3,000 – 7,000 |

Order-of-magnitude estimates are usually practiced by experienced engineers who have worked on similar projects in the past. They take minutes or hours to complete but the error in the estimate can be as high as 50%. Most engineers employed by operating companies usually perform level 2 and 3 studies. Such studies take days or weeks to complete using appropriate computer aids. The main objective of such studies is to evaluate alternatives and pinpoint the most cost-sensitive areas - the economic "hot-spots" - of a complex process.

The results of such analyses are used to plan future research and development and to generate project budgets. Level 4 and 5 studies are usually performed by engineering and construction companies that are hired to build new plants for promising new products that are at an advanced stage of development.

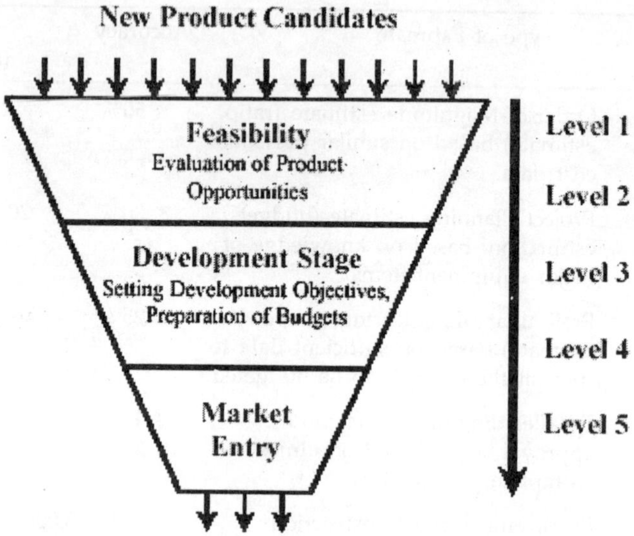


*Figure 1. Types of design estimates during the lifecycle of a product.*

It should be noted that opportunities for creative process design work are usually limited to preliminary studies. By the time detailed engineering work is initiated, a process is more than 80% fixed. Furthermore, the vast majority of important decisions for capital expenditures and product commercialisation are based on results of preliminary process design and cost analysis. This explains why it is so important for a new engineer to master the skills of preliminary process design and cost estimation. Environmental impact assessment is an activity closely related to process design and cost estimation.

Biochemical plants generate a wide range of liquid, solid, and gaseous waste streams that require treatment prior to discharge. The cost associated with waste treatment and disposal has skyrocketed in recent years due to increasingly stricter environmental regulations. This cost can be reduced through minimisation of waste generation at the source. However, generation of waste from a chemical or biochemical process is dependent upon the process design and the manner in which the process is operated.

Thus, reducing waste in an industrial process requires intimate knowledge of the process technology, in contrast to waste treatment which essentially is an add-on at the end of the process. In addition, minimisation of waste generation must be considered by process engineers at the early stages of process development. Once a process has undergone significant development it is difficult and costly to make major changes.

Furthermore, regulatory constraints that are unique to the pharmaceutical industry restrict process modifications once clinical efficacy of the drug is established. These are only some of the reasons that process synthesis must be considered not only during, but before, the selection of unit operations for individual steps.

## Bioseparation Processes Synthesis

The development of a flowsheet for the recovery and purification of a biological product is a creative process that draws on the experience and imagination of the engineer. Attempts have been made to capture that experience on the computer in the form of expert systems and automate to some extent the process synthesis tasks. Experienced engineers heavily rely on certain rules of thumb, also known as heuristics, for putting together the skeleton of a recovery and purification process. A few such heuristics follow:

1) Remove the most plentiful impurities first.
2) Remove the easiest-to-remove impurities first.
3) Make the most difficult and expensive separations last.
4) Select processes that make use of the greatest differences in the properties of the product and its impurities.
5) Select and sequence processes that exploit different separation driving forces.

Figure 2 provides a generalised structure for putting together an initial block diagram representation of a recovery process. For each product category (intracellular or extracellular) several branches exist in the main pathway. Selection among the branches and alternative unit operations is based on the properties of the product, the properties of the impurities, and the properties of the producing microorganisms, cells or tissues.

Bioprocess synthesis thus consists of sequencing steps according to the five heuristics and the structure of Figure 2. The majority of bioprocesses, especially those employed in the production of highvalue, low-volume products operate in batch mode. Continuous bioseparation processes are utilised in the production of commodity biochemicals, such as organic acids and ethanol.

### *Primary Recovery Stages*

Primary recovery comprises the first steps of downstream processing where some purification and broth volume reduction occurs. According to Figure 2, the selection of the first step depends on whether the product is intracellular (remains inside the microorganism after its expression) or extracellular (secreted into the solution). Almost all low molecular weight and many high molecular weight bioproducts are extracellular. Their recovery and

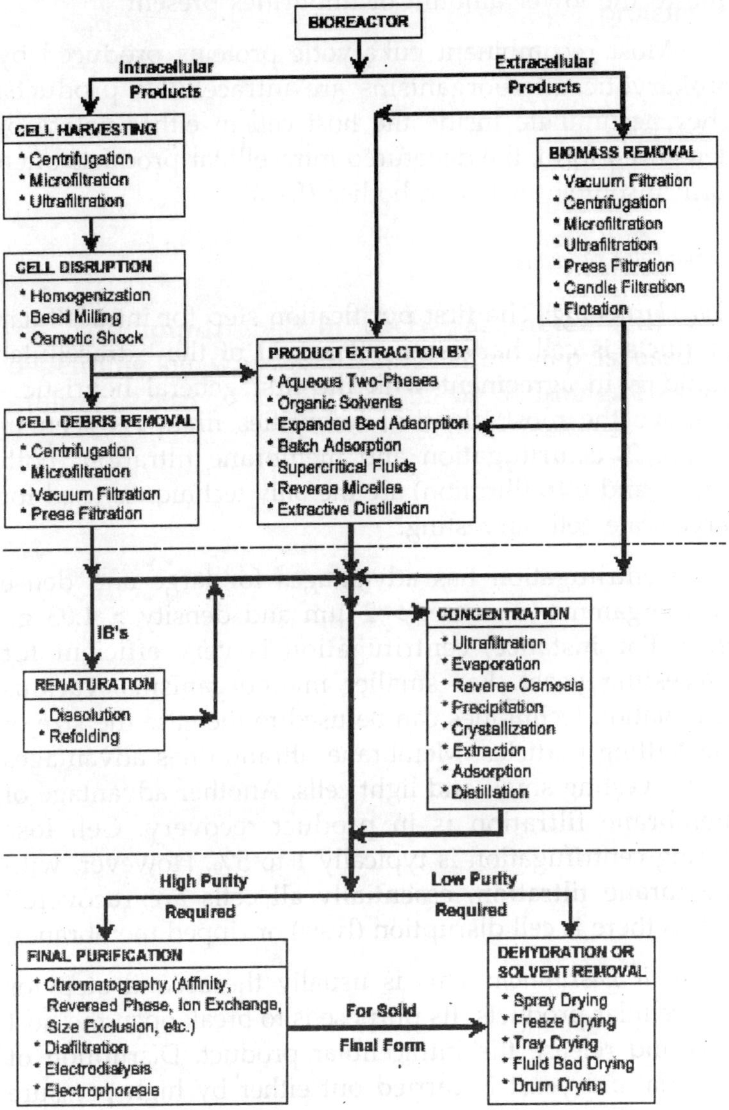


*Figure 2. Generalized block diagram of downstream processing.*

purification is easier compared to intracellular products due to the lower amount of impurities present.

Most recombinant eukaryotic proteins produced by prokaryotic microorganisms are intracellular products. They accumulate inside the host cell in either native or denatured form; the denatured intracellular products often form insoluble inclusion bodies (IB's).

### Intracellular products

*Cell Harvesting*: The first purification step for intracellular products is cell harvesting. Removal of the extracellular liquid is in agreement with the first general heuristic - Remove the most plentiful impurities first. As seen in Figure 2, centrifugation and membrane filtration (both micro- and ultrafiltration) are the only techniques used for large-scale cell harvesting.

Centrifugation has advantages for large and dense microorganims (diameter > 2 μm and density > 1.03 g/cm$^3$). For instance, centrifugation is very efficient for harvesting yeast. For smaller microorganisms, various coagulation techniques can be used to increase the size of the settling particles. Membrane filtration has advantages for harvesting small and light cells. Another advantage of membrane filtration is in product recovery. Cell loss during centrifugation is typically 1 to 5%. However, with membrane filtration, essentially all cells are recovered unless there is cell disruption (lysis) or ripped membranes.

*Cell Disruption*: This is usually the second step for intracellular products. Its purpose is to break open the host cells and release the intracellular product. Disruption of bacteria and yeast is carried out either by high pressure homogenisers or bead mills. For large capacities (several m$^3$/h) only high pressure homogenisers are practical. Osmotic shock is often used for release of periplasmic products that accumulate between the cell membrane and the cell wall Prior to disruption the concentrate is often diluted (by 5-10%) with a "lysis buffer" to create conditions that minimise product denaturation upon release from the cell. For hard-to-disrupt microorganisms, multiple homogeniser passes at 500-1000 bar are required.

Multiple passes are also required if the product forms inclusion bodies.This allows the inclusion bodies to be released, and also breaks the cell debris into very small particles, which facilitates the separation of inclusion bodies from cell debris further downstream. Some product protein degradation occurs during cell disruption due to high shear and oxidation.

*Removal of cell debris*: The cell debris that is generated by cell disruption is usually removed by centrifugation or microfiltration. Other options include rotary vacuum filtration, press filtration, depth filtration, extraction, and expanded bed adsorption (EBA) chromatography.

— *Soluble Product*. When the product is soluble, it is recovered during cell debris removal either in the light phase of a centrifuge or in the permeate stream of a filter. Centrifuges efficiently separate only fairly large particles of cell debris. Therefore, when a centrifuge is used for cell debris removal, a polishing filtration step must follow to remove small debris particles which might otherwise cause severe problems in processes downstream such as chromatography.

  Various types of filters (e.g., depth, press, candle, rotary vacuum, membrane microfilters, etc.) can be used for polishing. Alternatively, these filters can be used for cell debris removal without a centrifugation step preceding them. It is very difficult to predict a priori which filter performs best for a specific product. When microfilters are used for cell debris removal, some degree of diafiltration is required to achieve an acceptable product recovery yield.

— *Insoluble Product*: When the product is insoluble and forms inclusion bodies, it must first be separated from the cell debris particles, then dissolved and refolded. Fortunately, inclusion bodies usually have a large diameter (0.3 - 1.0 mm) and high density (1.3 - 1.5 g/cm$^3$) and can be separated from cell debris with a disk-stack centrifuge. The inclusion bodies are recovered in the heavy phase of the centrifuge while most cell debris particles remain in the light phase. The heavy phase is usually resuspended and recentrifuged 2-3 times to reach a high degree of inclusion-body purity.

Resuspension in a solution of a detergent and/or a low concentration of a chaotropic agent is often practiced to facilitate the removal of other contaminants. The pH and the ionic strength of the solution are adjusted to reduce the hydrophobicity of the cell debris particles and enhance their removal in the light phase. Final product purity exceeding 70% is quite common.

*Product Extraction / Adsorption*: Product separation from cell debris can also be carried out by extraction and/or adsorption. Organic solvents are commonly used as extractants for low molecular weight products, such as various antibiotics. Aqueous two-phase systems have found applications for recovery of proteins. The criteria for extractant selection are: the partition coefficient of the product should be higher than the partition coefficient of the contaminants; the extractant should not degrade the product; it should not be expensive; and it should be easy to recover or dispose of.

In addition, product separation from debris and simultaneous concentration can be achieved by adsorptive techniques. Various types of adsorbents (e.g., ion exchange, reverse phase, affinity, etc.) can be used. This type of purification requires the disrupted cells and product to be mixed in a stirred tank with an adsorbent. A washing step, where most of the cell debris particles and contaminants are washed out, follows product adsorption.

More recently, expanded bed adsorption (EBA) chromatography has shown promise for separating proteins from cell debris particles. The feed is pumped upwards through an expanded bed. Target proteins are bound to the adsorbent while cell debris and other contaminants pass through. A washing step removes all weakly retained material. An elution step follows that releases and further purifies the product.

#### Extracellular products

*Biomass Removal*: In agreement with the second generic heuristic, remove the easiest-toremove impurities first, biomass removal is usually the first step of downstream processing of extracellular products. This step can be accomplished by using one (or more) of the following unit operations: rotary vacuum filtration, disk-stack or decanter centrifugation, press filtration, microfiltration, ultrafiltration, flotation, etc. Since each unit operation has advantages and disadvantages for different products and microorganisms, the selection of the best unit operation(s) for a given system can be difficult.

— *Rotary vacuum filtration*, especially with precoat, is the classical widely used method for removal of mycelial organisms. Rotary vacuum filters can operate continuously for long periods of time. In addition, the filtrate flux in these units is usually higher than 200 $L/m^2$-h and may reach 1,000 L/ $m^2$-h. The most important disadvantage of this type of unit is the problem with the disposal of the mixture of filter-aid and biomass. Filter-aid is added in equal or higher amounts than biomass. Stringent environmental laws have made it costly to dispose of such solid materials.

Therefore, if the disposal cost of filter aid is relatively high where a new plant is going to be built, alternative unit operations should be considered for biomass separation. However, if the disposal cost of filter aid is relatively low, a rotary vacuum filter is a good choice. The citric acid process, offers an example where rotary vacuum filtration is used for biomass removal.

— *Centrifugation*: Disk-stack and decanter centrifuges are frequently used at large scale. Of the two, disk-stack centrifuges operate at higher rotational speeds and remove smaller and lighter microorganisms. However, with the use of coagulating agents, the decanter centrifuge performance improves, and choosing between the two types becomes more difficult. It appears that the only criterion being used to choose disk-stack as opposed to decanter is the ability to remove small, light microorganisms.

  Centrifugation does not require filter aid, which is a significant advantage compared to rotary vacuum filtration. In general, the centrifuge paste contains 40-60% v/v extracellular liquid. In order to recover the product dissolved in that liquid, the paste is usually washed and re-centrifuged.

— *Membrane filtration*: With membrane filters (micro- and ultrafilters) the extracellular product passes through the membrane while biomass and other particulate components remain in the concentrate. Concentration is usually followed by diafiltration to increase the product recovery yield. Membrane filters are used for biomass removal mainly in recovery of low molecular weight products, such as antibiotics from mycelia. For high molecular weight products, applications are limited to cases where the amount of solids is rather small as in cell culture.

### *Intermediate Recovery Stages*

The primary recovery stages just described are followed by the intermediate stages, where the product is concentrated and further purified. If the product is soluble, product concentration is usually the first step. If the product is denatured and insoluble, first it is dissolved and refolded and then it is concentrated and purified.

*Product Concentration*. After primary separation, the product is usually in a dilute solution. Volume reduction by concentration is in agreement with heuristics 1 and 2. Common concentration options include ultrafiltration, reverse osmosis, evaporation, adsorption, precipitation, extraction, and distillation.

— *Ultrafiltration*: is used extensively for protein solution concentration. The molecular weight cut-off of the membrane is selected to retain the product while allowing undesirable impurities (mainly low molecular weight solutes) to pass through the membrane. The low operating temperature and the purification achieved along with concentration are some of the advantages of ultrafiltration over evaporation. The typical operating trans-membrane pressure is 2-5 bar and the average flux is 20-50 $L/m^2$-h.

— *Reverse Osmosis* filters employ membranes with smaller pore sizes and are used for concentrating medium to low molecular weight products (e.g., antibiotics, certain amino acids, etc.).

— *Evaporation*: Thin-film rotating evaporators can operate at relatively low temperatures (40-50°C) under vacuum. These units compete in the market with ultrafiltration and reverse osmosis for concentrating both low and high molecular weight compounds. One disadvantage of evaporation compared to ultrafiltration is the lack of any purification during concentration. Advantages include the ability to concentrate to a higher final solids concentration and the ability to handle large throughputs.

— *Precipitation* is often used for concentration and purification. Blood protein fractionation and citric acid production constitute typical applications. Addition of salts, solvents, and polymers and changes in pH, ionic strength, and temperature are commonly used to selectively precipitate compounds of interest. Precipitation often follows an extraction carried out by a polymer/salt (e.g., PEG and potassium phosphate) aqueous two-phase system. When the product is recovered in the polymerrich phase, precipitation is accomplished by addition of more polymer. It is important for economic reasons to recover and recycle the precipitating materials. Precipitation is also used to remove contaminants, i.e. nucleic acids, by adding MnSO4 and streptomycin sulfate.

— *Distillation* is used for concentrating and purifying organic solvents, such as ethanol, acetic acid, etc.

*Product Renaturation*: Eukaryotic proteins produced by prokaryotic microorganisms often form insoluble inclusion bodies (IB's) in the host cell. Inclusion bodies can be dissolved rapidly using solutions of strong chaotropes, such as 6 M guanidine hydrochloride or urea, in the presence of a reducing agent, such as 0.5 M 2-mercaptoethanol or 50 mM dithiothrietol. The dissolved protein is then allowed to refold to its native conformation by removing the chaotropic agents through diafiltration or chromatography and diluting the solution down to total protein concentration of 20-50 mg/L.

Dilution is necessary for minimising intermolecular interactions, which occur during product refolding and can lead to product inactivation. Addition of small amounts of reduced glutathione (2-5 mM) and oxidised glutathione (1-2 mM) and incubation at 35-40 °C for 5-10 hours completes the refolding process. Thus, when choosing an upstream process that forms IB's consideration must be given to the large volumes, and hence large waste streams, that are produced.

### Final Purification Stages

The final purification steps are dependent on the required final product purity. Pharmaceutical products require high purity while industrial products require lower purity. For products of relatively low purity, such as detergent enzymes, the final purification step is dehydration or more generally a solvent removal step. For high purity products, the final purification stages usually involve a combination of chromatographic and filtration steps. If the final product is required in solid form, then, a dehydration or solvent removal step follows.

*Chromatography* is typically done later in a process in agreement with the third generic heuristic, make the most difficult and expensive separations last. With the previous separation steps, a large fraction of contaminants is removed, thereby reducing the volume of material that needs to be treated further. In fact, a 50-100 fold volumetric reduction is quite common for highvalue biological products, resulting in a protein content of 1-5% w/v in the feed stream to chromatographic units.

Recent advances in expanded bed adsorption (EBA) chromatography promise to position chromatographic steps in the primary recovery stages. EBA chromatography units can be used to capture, concentrate, and purify product directly from fermentation broth that contains whole cells, cell debris and other particulate components. Consequently, EBA chromatography has the potential to eliminate some of the typical primary recovery steps, such as biomass and cell debris removal, product concentration, etc.

A sequence of chromatographic steps is usually required to achieve the desired final product purity, and the fourth and fifth generic heuristics are good guides for selecting and sequencing such steps. For instance, according to the fifth heuristic, an ion exchange step should not be followed by another step of the same type. Instead, it should be followed by a reverse phase, affinity or any other chromatography step that takes advantage of a different separation driving force. Membrane filtration steps are commonly employed between chromatographic steps to exchange buffers and concentrate the dilute product solutions.

*Dehydration or Solvent Removal* is achieved with dryers. Spray, fluid bed, and tray dryers are used when products can withstand temperatures of 50-100 °C. Freeze dryers are used for products that degrade at high temperatures. Freeze dryers require high capital expenditures and should be avoided if possible.

### Analysis of Process Synthesis

The flowsheets put together during process synthesis must be analysed and compared on the basis of capital investment, manufacturing cost, environmental impact, and other criteria in order to decide which ideas to consider further. The estimation of capital investment and manufacturing cost is based on the results of material and energy balances and equipment sizing. Although these basic chemical engineering calculations can be done on paper, it is highly desirable to use more sophisticated computer aids, such as spreadsheets or process simulators. Use of computer tools allows the process design team to quickly and accurately redo the entire series of calculations for a different set of assumptions and other input data.

### *Spreadsheets*

Spreadsheet applications, such as Microsoft Excel, Lotus 1-2-3, and Corel Quattro Pro have become as easy to use as word processors and graphics packages. In its simplest form, a spreadsheet is an electronic piece of paper with empty boxes, known as cells. The user can enter data in those cells, perform calculations, and generate results. Results from spreadsheets can be easily plotted in a variety of graphs.

### *Process Simulators*

Process simulators are software tools that enable the user to readily represent and analyse integrated processes. They have been in use in the petrochemical industries since the early 1960's. Established simulators for the petrochemical industries include: Aspen Plus (from Aspen Technology, Inc.), ChemCAD (from Chemstations, Inc.), HYSYS (from Hyprotech, Ltd./AEA Engineering Software), and PRO/II (from Simulation Sciences, Inc.). Devclopment of simulators specific to biochemical processes began in the mid 1980's.

BioProcess Simulator (BPS) (from Aspen Technology, Inc.) was the first tool of this type. For a given flowsheet, BPS used to carry out material and energy balances, estimate the size and cost of equipment, and perform economic evaluation. BPS has had limited commercial success because it was designed as an extension of Aspen Plus, an inherently steady-state simulator, and could not satisfactorily represent batch biochemical processes, which normally operate in batch mode.

BioPro Designer, the second product of this category, was initially developed at the Biotechnology Process Engineering Center (BPEC) of MIT. With a license to this technology, Intelligen, Inc. completed the development of

BioPro Designer and commercialised it, first for the Apple Macintosh and later for MS Windows. SuperPro Designer, an extension of BioPro, was created to address other related industries (e.g., synthetic pharmaceuticals, agrochemicals, food processes, etc.) as well as water purification and end-ofpipe treatment processes.

SuperPro handles material and energy balances, equipment sizing and costing, economic evaluation, environmental impact assessment, process scheduling, and debottlenecking of batch and continuous processes. Biotechnology Design Simulator (BDS), the third tool of this family, was developed by Life Sciences International. BDS runs on top of Gensym's G2 system and focuses on scheduling of batch operations and resource utilisation as a function of time.

BATCHES from Batch Process Technologies is a batch process simulator that has found applications in pharmaceuticals, biochemicals, and food processing. It is especially useful for fitting a new process into an existing facility and analysing resource demand as a function of time. More recently, Aspen Technology and Hyprotech have introduced Batch Plus and BDK, respectively. Both mainly target synthetic pharmaceutical and specialty chemical processes, but it is expected that future versions of these tools will be able to handle biochemical processes as well. Minimum requirements for a biochemical process simulator are the ability to handle batch as well as continuous processes and the ability to model the unit operations that are specific to bioprocessing. Due to its ability to satisfy these requirements, SuperPro Designer will be used to illustrate the role of such tools in bioprocess design.

To model an integrated process on the computer using a simulator, you must start by developing a flowsheet that represents the overall process. The user initialises the flowsheet by registering (selecting from the component database) the various materials that are used in the process and specifying operating conditions and performance parameters for the various operations. Most biochemical processes operate in batch or semi-continuous mode. This is in contrast to continuous operation, which is typical in the petrochemical and other industries that handle large throughputs.

In continuous operations, a piece of equipment performs the same action all the time (which is consistent with the notion of unit operations). In batch processing, on the other hand, a piece of equipment goes through a cycle of operations. For instance, a typical chromatography cycle includes equilibration, loading, washing, elution, and regeneration. In SuperPro Designer v4.0, the set of operations that comprise a processing step is called a "unit procedure" (as opposed to a "unit operation"). Each unit procedure contains individual tasks (e.g., equilibration, loading, etc.) called operations.

A unit procedure is represented on the screen with a single equipment icon. In essence, a unit procedure is the recipe of a processing step that describes the sequence of actions required to complete that step. The significance of the unit procedure is that it enables the user to describe and model the various activities of batch processing steps in detail.

For every operation within a unit procedure, SuperPro includes a mathematical model that performs material and energy balance calculations. Based on the material balances, SuperPro performs equipment-sizing calculations similar to some of the homework problems in the book. If multiple operations within a unit procedure dictate different sizes for a certain piece of equipment, the software reconciles the different demands and selects an equipment size that is appropriate for all operations.

In other words, the equipment is sized so that it is large enough that it will not be overfilled during any operation, but it is no larger than necessary. In addition, the software checks to ensure that the vessel contents will not fall below a user-specified minimum volume (e.g., a minimum stir volume) for applicable operations. Before any simulation calculations can be done, the user must initialise the various operations by specifying operating conditions and performance parameters through appropriate dialog windows.

Through the labor, etc. tab of the same dialog window, the user provides information about labor requirement during this operation. Through the scheduling tab one specifies the start time of this operation relative to the start or end time of another operation in the same procedure, or relative to an operation in another procedure. After initialisation of the operations, the simulator performs material and energy balances for the entire process and estimates the required sizes of equipment.

Optionally, the simulator may be used to carry out cost analysis and economic evaluation calculations. Other tasks that can be handled by process simulators include process scheduling, environmental impact assessment, debottlenecking, and throughput analysis. Issues of process scheduling and environmental impact assessment will be addressed in the examples section.

In throughout analysis and debottlenecking, the engineer analyses the capacity and time utilisation of equipment and resources (e.g., utilities, labor, raw materials) and tries to identify opportunities for increasing throughput with the minimum possible capital investment. Having developed a good model using a process simulator or a spreadsheet, the user may begin experimenting on the computer with alternative process setups and operating conditions. This has the potential of reducing the costly and time-consuming laboratory and pilot plant effort. Please be aware, however, that the GIGO (garbage-in, garbage-out) principle applies to all computer models.

More specifically, if some of your assumptions and input data are incorrect, so will be the outcome of the simulation. Consequently, a certain validation of the model is necessary. In its simplest form, a review of the results by an experienced engineer can play the role of validation.

The use of SuperPro Designer is illustrated to analyze and evaluate the production of three biological products. The first example analyses the production of citric acid, a commodity organic acid. The second deals with the production of recombinant human insulin, the first commercial product of modern biotechnology. The third example focuses on the production of monoclonal antibodies (MAb's) using mammalian cells cultured in stirred-tank bioreactors. The generation of the flowsheets for the production of all three products was based on information available in the patent and technical literature combined with our engineering judgment and experience with other biological products.

### Citric Acid Production

A number of organic acids are produced via fermentation. Of these, citric acid is produced in the largest amount. Citric acid is marketed as citric acid-1-hydrate or as anhydrous citric acid. The majority of citric acid (more than 60%) is used in the food and beverage industries to preserve and enhance flavor. In the chemical industries (25-30% of total), citric acid is used for the treatment of textiles, as a softener, as an antifoam agent, etc.

In the pharmaceutical industry (10% of total), iron citrate is used as a source of iron and citric acid is used as a preservative for stored blood, tablets, ointments, and in cosmetic preparations. Citric acid is being used more and more in the detergent industry as a replacement for polyphosphates. Citric acid was first recovered in 1869 in England from calcium citrate, which was obtained from lemon juice. Its production by filamentous fungi has been known since 1893.

The first production via surface culture fermentation was initiated in 1923. Production using stirredtank fermentors began in the 1930's and it is presently the preferred method for large scale manufacturing. The plant analysed in this example produces around 10,000,000 kg of crystal citric acid per year, which represents approximately 2.5% of the current world demand.

*Upstream section*: Molasses, the carbon source of fermentation, is diluted from about 50% fermentable sugars content to 20% with water in a blending tank. Suspended particulate material is removed by filtration. Metal ions, particularly iron, are removed by an ion-exchange chromatography column. The purified raw material solution is then heat-sterilised. Nutrients (i.e., sources of ammonium, potassium, phosphorous, magnesium, copper, and zinc) are dissolved in water and heat-sterilised. The fermentation cycle is 7 days and the production is handled by 7 fermentors that operate in staggered mode. Since the plant operates around the clock, one fermentation cycle is initiated daily and another one is completed daily.

Each fermentor has a vessel volume of 260 $m^3$ and handles broth of around 207.4 $m^3$. Each production fermentor is preceded by two smaller seed fermentors, which are not shown in the flowsheet. A pure culture of the mold Aspergillus niger is used to inoculate the smallest seed fermentor. When optimum growth of mycelium is reached, the contents of the seed fermentor are transferred to the next stage fermentor, which is approximately ten times larger. Similarly, this larger seed fermentor inoculates the production fermentor with about 10% volume of actively growing mycelium broth.

Air is supplied by a compressor at a rate that gradually increases from 0.15 to 1.0 VVM (volume of air per volume of liquid per minute). Cooling water removes the heat produced by the exothermic process (2,990 kcal/kg of citric acid formed) and maintains the temperature at 28°C. The fermented broth is discharged into the holding tank, which acts as a buffer tank between the batch upstream section and the continuous downstream section.

*Dowstream section*: Purification starts with the removal of biomass by a rotary vacuum filter (RVF). The clarified fermentation liquor flows to an agitated reaction vessel. Approximately 1 part of hydrated lime, $Ca(OH)_2$, for every 2 parts of liquor is slowly added to precipitate calcium citrate. The lime solution must be very low in magnesium content if losses due to the relatively soluble magnesium citrate are to be avoided. Calcium citrate is separated by a second rotary vacuum filter and the citrate-free filtrate is disposed of.

The calcium citrate cake is sent to another agitated reaction vessel where it is acidified with dilute sulfuric acid to form a precipitate of calcium sulfate (gypsum). A third filter removes the precipitated gypsum and yields an impure citric acid solution. Careful control of pH and temperature of the precipitation steps is important for maximising the yield of citric acid. The resulting solution is concentrated and crystallised using a continuous evaporator / crystalliser.

The crystals formed are separated by filtration and dried in a rotary dryer. If the final product is required in high purity, treatment with activated carbon may precede crystallisation to remove colorants. Ion exchange is sometimes used to remove metal ions and other ionic species.

"CA Crystal" stands for crystalline citric acid and represents the final product. Glucose represents the fermentable carbohydrates in molasses (50% w/w). Note the large amounts of $Ca(OH)_2$ and sulfuric acid consumed and gypsum (calcium sulfate) generated. Their quantities depend on the chemistry of the recovery process and cannot be reduced without changing the recovery technology. Since this gypsum is contaminated with biomass, it has little or no commercial value. A disposal cost of $50/ton (metric) was assumed in this example. The large amount of wastewater is also worth noting.

### Human Insulin Production

Insulin facilitates the metabolism of carbohydrates and is essential for the supply of energy to the cells of the body. Impaired insulin production leads to the disease diabetes mellitus, which is the third largest cause of death in industrialised countries after cardiovascular diseases and cancer. Human insulin is a polypeptide consisting of 51 amino acids arranged in two chains: A with 21 amino acids, and B consisting of 30 amino acids.

The A and B chains are connected by two disulfide bonds. Human insulin has a molecular weight of 5,734 and an isoelectric point of 5.4. Human insulin can be produced by four different methods:

— Extraction from human pancreas.

— Chemical synthesis via individual amino acids.

— Conversion of pork insulin or "semisynthesis."

— Fermentation of genetically engineered microorganisms.

Extraction from the human pancreas cannot be practiced due to the limited availability of raw material. Total synthesis, while technically feasible, is not economically viable due to the very low yield. Production based on pork insulin, also known as "semisynthesis," transforms the porcine insulin (which differs only in one amino acid) molecule into an exact replica of the human insulin molecule by substituting the amino acid threonine for alanine in the G-30 position.

However, this option is also quite expensive because it requires the collection and processing of large amounts of porcine pancreases. In addition, its supply is limited by the availability of porcine pancreas. At least three alternative technologies have been developed for producing human insulin based on fermentation and utilising recombinant DNA technology.

— *Two-chain Method*: This was the first successful technique of biosynthetic human insulin (BHI) production based on recombinant DNA technology. Each insulin chain is produced as a b-galactosidase fusion protein in Escherichia coli forming inclusion bodies. The two peptide chains are recovered from the inclusion bodies, purified, and combined to yield human insulin. Later, the b-galactosidase operon was replaced with the tryptophan (Trp) operon, resulting in a substantial yield increase.

— *Proinsulin method (intracellular)*: This method eliminates the need for two different fermentation and purification trains that the previous option requires. In this case, intact proinsulin is produced instead of the separate A and B chains. Figure 3 shows the key transformation steps. The E. coli cells overproduce Trp-LE'-Met-proinsulin in the form of inclusion bodies, which are recovered and solubilised. Proinsulin is released by cleaving the methionine linker using

CNBr. The proinsulin chain is subjected to a folding process to allow intermolecular disulfide bonds to form, and the C peptide is then cleaved with enzymes to yield human insulin. A number of chromatography and membrane filtration steps are utilised to purify the product.

— *Proinsulin method (secreted)*: Secretion simplifies product isolation and purification. The precursor contains the correct disulfide bridges and is therefore identical to those of insulin. It is converted to human insulin by transpeptidation in organic solvent in the presence of a threonine ester and trypsin followed by de-esterification. Another advantage of this technology is the ability to reuse the cells by employing a continuous bioreactor-cell separator loop.

#### Design Basis Market Analysis

The current world demand for insulin is in the range of 15,000 to 25,000 kg per year and it is growing at an annual rate of 5% to 6%. The plant analysed in this example has a capacity of around 1,800 kg of purified biosynthetic human insulin (BHI) per year. This is a relatively large plant for producing polypeptide-based biopharmaceuticals. The plant operates around the clock for 330 days a year. A new batch is initiated every 48 hours resulting in 160 batches per year. The fermentation broth volume per batch is approximately 37.5 $m^3$.

The entire flowsheet for the production of BHI is divided into four sections: 1) Fermentation, 2) Primary Recovery, 3) Reactions, and 4) Final Purification. (a "section" in SuperPro is simply a set of unit procedures). If you open the computer file ("insulin.spf") using SuperPro, you will see that the unit procedures in each section have their own distinctive color (blue, green, purple, and black for Fermentation, Primary Recovery, Reactions, and Final Purification, respectively).

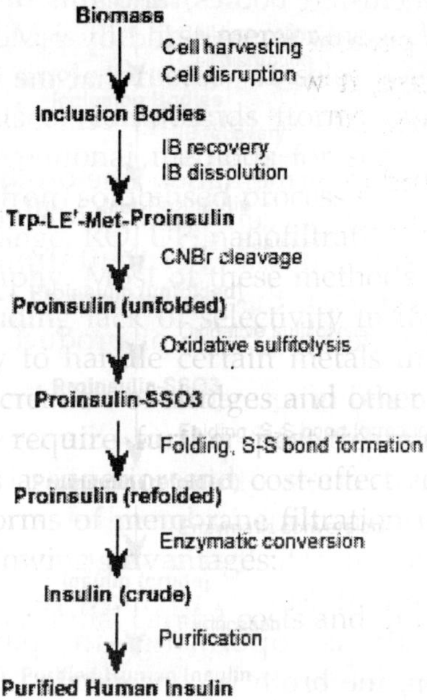


*Figure 3. Human insulin from proinsulin fusion protein.*

Fermentation media are prepared in a stainless steel tank and sterilised in a continuous heat steriliser. The axial compressor and the absolute filter provide sterile air and ammonia to the fermentor at an average rate of 0.5 VVM. A two-step seed fermentor train (not shown in the flowsheet) is used to inoculate the 50 $m^3$ production fermentor with transformed E. coli cells. These cells are used to produce the Trp-LE′-MET-proinsulin precursor of insulin, which is retained in the cellular biomass.

The fermentation time in the production fermentor is about 18 hours, and the fermentation temperature is 37 °C. The final concentration of E. coli in the production fermentor is about 30 g/liter (dry cell weight). The Trp operon is turned on when the E. coli fermentation runs out of tryptophan. The chimeric protein Trp-LE′-MET-proinsulin accumulates intracellularly as insoluble aggregates (inclusion bodies) and this decreases the rate at which the protein is degraded by proteolytic enzymes. In the base case, it was assumed that the inclusion bodies (IB's) constitute 20% of total dry cell mass.

At the end of fermentation, the broth is cooled down to 10°C to minimise cell lysis. After completing each processing step in the Fermentation Section (and subsequent sections), the equipment is washed in order to prepare for the next batch of product.

After the end of fermentation, the broth is transferred into a surge tank which isolates the upstream from the downstream section of the plant. Three disk stack centrifuges operating in parallel are used for cell harvesting. Please note that a single unit procedure icon on the screen of SuperPro may represent multiple equipment items operating in parallel. During centrifugation, the broth is concentrated from 37,000 L to 9,165 L, and most of the extracellular impurities are removed. The cell recovery yield is 98%. The cell sludge is diluted with an equal volume of buffer solution (buffer composition: 96.4% w/w WFI (water for injection), 0.7% EDTA, and 2.9% TRIS-Base ) using a blending tank.

The buffer facilitates the separation of the cell debris particles from inclusion bodies. Next, a high pressure homogeniser is used to break the cells and release the inclusion bodies. The broth undergoes three passes under a pressure drop of 800 bar. The exit temperature is maintained at around 10°C. The same centrifuges as before are then used for inclusion body recovery. The reuse of these centrifuges can be seen by the fact that procedures P-9 and P-13 have the same equipment name, DS-101.

The IB's are recovered in the heavy phase (with a yield of 98%) while most of the cell debris particles remain in the light phase. This is possible because the density (1.3 $g/cm^3$) and size (diameter ca. 1 micron) of the IB's are significantly greater than that of the cell debris particles. The IB sludge, which contains approximately 20% solids w/w, is washed with WFI containing 0.66% w/w Triton-X-100 detergent and recentrifuged using the same centrifuges as before. The detergent solution facilitates purification (dissociation of debris and soluble proteins from inclusion bodies). The exit temperature is maintained at 10 °C. The slurry volume at the end of the primary recovery section is around 1,400 L.

*Inclusion Body Solubilization*: The inclusion body suspension is transferred to a glasslined reaction tank (V-103) and is mixed with urea and 2-mercaptoethanol to final concentrations of 300 g/L (5 M) and 40 g/L, respectively. Urea is a chaotropic agent that dissolves the denatured protein in the inclusion bodies and 2-mercaptoethanol is a reductant that reduces disulfide bonds. A reaction time of 8 hours is required to reach a solubilisation yield of 95%.

The inclusion bodies are composed of 80% w/w Trp-LE′-Met-proinsulin, with the remainder being other (contaminant) proteins. After the end of the solubilisation reaction, urea and 2-mercaptoethanol are replaced with WFI and the solution is concentrated using a diafiltration unit (DF-101). This operation is performed in 6 hours with a recovery yield of 98%. All remaining fine particles (biomass, debris, and inclusion bodies) are removed using a polishing dead-end filter. This polishing filter protects the chromatographic units that are used further downstream. The solution volume at this point is around 5,200 L.

*CNBr cleavage*: The chimeric protein is cleaved with CNBr (cyanogen bromide) into the signal sequence Trp-LE′-Met, which contains 121 amino acids, and the denatured proinsulin (82 amino acids) in the same reactor that was used for IB solubilisation. The reaction is carried out in a 70% formic acid solution containing 30-fold molar excess CNBr (stoichiometrically, one mole of CNBr is required per mole of Trp-LE'-Met-proinsulin). The reaction takes 12 hours at 20°C and reaches a yield of 95%.

The mass of the released proinsulin is approximately 30% of the mass of Trp-LE'-Met-proinsulin. A small amount of cyanide gas is formed as a by-product of the cleavage reaction. The formic acid, unreacted CNBr, and generated cyanide gas are removed by applying vacuum and raising the temperature to around 35 °C (the boiling point of CNBr). This operation is carried out in a rotary vacuum evaporator and takes 1 hour. Since cyanide gas is toxic, all air exhausted from the vessels is scrubbed with a solution of hypochlorite, which is prepared and maintained in situ.

*Sulfitolysis*: Sulfitolysis of the denatured proinsulin takes place in a reaction tank under alkaline conditions. This operation is designed to unfold proinsulin, break any disulfide bonds, and add SO3 moieties to all sulfur residues on the cysteines. The product of interest is human proinsulin(S-SO3-)6 (protein-S-sulfonate). The sulfitolysis step is necessary for two reasons:

1) the proinsulin is probably not folded in the correct configuration when expressed in E. coli as part of a fusion protein, and

2) the cyanogen bromide treatment tends to break existing disulfide bonds.

The final sulfitolysis mixture contains 50% w/w guanidineoHCl (6 M), 0.35% ammonium bicarbonate ($NH_4HCO_3$), 3% $Na_2SO_3$ and 1.5% $Na_2O_6S_4$. A reaction time of 12 hours is required to reach a yield of 95%. The presence of the denaturing reagent (guanidineoHCl) prevents refolding and cross-folding of the same protein molecule onto itself or two separate protein molecules onto each other. Urea may also be used as a denaturing reagent. After the completion of the sulfitolysis reaction, the sulfitolysis solution is exchanged with WFI to a final guanidineoHCl concentration of 20% w/w.

The human proinsulin(S-$SO_{3-}$)$_6$ is then chromatographically purified using three ion-exchange columns operating in parallel. Each column has a diameter of 140 cm and a bed height of 25 cm. A cation exchange resin is used operating at pH 4.0. The eluant solution contains: 69.5 % w/w WFI, 29% urea, and 1.5% NaCl. Urea, a denaturing agent, is used to prevent incorrect refolding and crossfolding of proinsulin (S-$SO_{3-}$)$_6$. The following operating assumptions were made:

1) the column is equilibrated for 30 minutes prior to loading,
2) the total resin binding capacity is 20 mg/ml,
3) the eluant volume is equal to 5 column volumes (CV's),
4) the total volume of the solutions for column wash, regeneration and storage is 15 CV's, and
5) the protein of interest is recovered in 1.5 CV's of eluant buffer with a recovery yield of 90%.

*Refolding*: This operation catalyses the removal of the $SO_3$ moiety and then allows disulfide bond formation and correct refolding of the proinsulin to its native form. It takes place in a reaction tank. This process step involves treatment with mercaptoethanol, a reductant that facilitates the disulfide interchange reaction. It is added at a ratio of 1.5 mol of mercaptoethanol to 1 mol of $SO_3$.

Dilution to a proinsulin(S-$SO_{3-}$)$_6$ concentration of less than 1 g/L is required to prevent cross-folding of proinsulin molecules. The reaction is carried out at 8°C for 12 hours and reaches a yield of 85%. After completion of the refolding step, the refolding reagents are replaced with WFI and the protein solution is concentrated using a diafiltration unit which has a product recovery yield of 95% (5% of the protein denatures). The volume of the solution at this point is around 5,000 L.

Next, the human proinsulin is chromatographically purified in a hydrophobic interaction chromatography column. The following operating assumptions were made: (1) the column is equilibrated for 30 minutes prior to loading, (2) the total resin binding capacity is 20 mg/ml, (3) the eluant volume is equal to 6 column volumes (CV's), (4) the total volume of the solutions for column wash, regeneration and storage is 15 CV's, (5) the protein of interest is recovered in 1 CV of eluant buffer with a recovery yield of 90%, and (6) the material of a batch is handled in three cycles.

*Enzymatic conversion*: The removal of the C-peptide from human proinsulin is carried out enzymatically (using trypsin and carboxypeptidase B) in a reaction tank. Trypsin cleaves at the carboxy terminal of internal lysine and arginine residues, and carboxypeptidase B removes terminal amino acids. The amount of trypsin used is rate-limiting and allows intact human insulin to be formed. Carboxipeptidase is added to a final concentration of 4 mg/liter, while trypsin is added to a final concentration of 1 mg/liter. The reaction takes place at 30 oC for 4 hours and reaches a conversion yield of 95%. The volume of the solution at this point is around 4,300 L.

#### *Final purification section*

A purification sequence based on multimodal chromatography, which exploits differences in molecular charge, size, and hydrophobicity, is used to isolate biosynthetic human insulin. A description of all the purification steps follows.

The enzymatic conversion solution is exchanged with WFI and concentrated by a factor of 4 in a diafilter. An ion exchange column is used to purify the insulin solution. The following operating assumptions were made:

1) the column is equilibrated for 30 minutes prior to loading,
2) the total resin binding capacity is 20 mg/ml,
3) the eluant volume is equal to 8 CV's and the eluant is a 11.5 % w/w solution of NaCl in WFI,
4) the total volume of the solutions for column wash, regeneration and storage is 14 CV's,
5) the protein of interest is recovered in 1.5 CV of eluant buffer with a recovery yield of 95%, and
6) the material from each batch is handled in four cycles.

The liquid volume at this point is around 1,100 L. Next, the ion exchange eluant solution is exchanged with WFI in a diafilter and is concentrated by a factor of 2.0. A recovery yield of 98% was assumed for this step (2% denatures). The purification of the insulin solution proceeds with a reverse phase high-pressureliquid-chromatography step.

Detailed information on the use of RP-HPLC for insulin purification is available in the literature. Analytical studies with a variety of reversedphase systems have shown that an acidic mobile phase can provide excellent resolution of insulin from structurally similar insulin-like components. Minor modifications in the insulin molecule resulting in monodesamido formation at the 21st amino acid of the A chain, or derivatisation of amines via carbamoylation or formylation, result in insulin derivatives which have significantly increased retention.

Derivatives of this nature are typical of the kind of insulin-like components that are found in the charge stream going into the reversed-phase purification. The use of an acidic mobile phase results in elution of all the derivatives after the insulin peak, while the use of mildly alkaline pH results in derivatives eluted on either side of the parent insulin peak. An ideal pH for insulin purification is in the region of 3.0-4.0, since this pH range is far enough below the isoelectric pH of 5.4 to provide for good insulin solubility.

An eluant buffer with an acetic acid concentration of 0.25 M meets these operational criteria since it is compatible with the chromatography and provides good insulin solubility. A 90% insulin yield was assumed in the RP-HPLC step with the following operating conditions: (1) the column is equilibrated for 30 minutes prior to loading, (2) the total resin binding capacity is 15 mg/ml, (3) the column height is 25 cm, (4) the eluant volume is equal to 6 CVs and its composition is 25% w/w acetonitrile, 1.5% w/w acetic acid 73.5% w/w WFI, (5) the total volume of the solutions for column wash, equilibration, regeneration and storage is 6 CVs, and (5) the protein of interest is recovered in 1 CV of eluant buffer with a recovery yield of 90%.

The RP-HPLC buffer is exchanged with WFI and concentrated by a factor of 2.0 in a diafilter (DF-105) that has a product recovery yield of 98% (2% denatures). Purification is completed by a gel filtration chromatography column (C-105). The following operating assumptions were made: (1) the column is equilibrated for 30 minutes prior to loading, (2) the sample volume is equal to 5% of the column volume, (3) the eluant volume is equal to 4 CV's, (4) the total volume of the solutions for column wash, depyrogenation, stripping, and storage is 6 CV's, and (5) the protein of interest is recovered in 0.5 CV of eluant buffer with a recovery yield of 90%. The mobile phase is a solution of acetic acid.

Next, the same diafilter is used to concentrate the purified insulin solution by a factor of ten. The liquid volume at this point is around 500 L, which contains approximately 12.8 kg of insulin. This material is pumped into a jacketed and agitated reaction tank. Ammonium acetate and zinc chloride are added to the protein solution until each reaches a final concentration of 0.02 M. The pH is then adjusted to between 5.4 and 6.2. The crystallisation is carried out at 5 oC for 12 hours.

Insulin crystallises with zinc with the following stoichiometry: insulin6-Zn2. Step recovery on insulin is around 90%. The crystals are recovered with a basket centrifuge with a yield of 95%. Finally, the crystals are freeze-dried. The purity of the crystallised end-product is between 99.5 and 99.9 % measured by analytical high pressure liquid chromatography (HPLC). The overall recovery yield is around 32%.

In the base case, a new batch is initiated every 48 hours. Most of the equipment items, however, are utilised for less than 24 hours per batch. If the market demand for insulin grows, this provides the opportunity for increasing plant throughput without major capital expenditures. A realistic improvement is to initiate a batch every 24 hours. This will require a new fermentor of the same size whose operation will be staggered relative to the existing unit so that one fermentor is ready for harvesting every day.

Such a production change will also require additional equipment of the following types: (1) disk-stack centrifuges to reduce the occupancy of DS-101 to less than 24 hours; (2) two reaction tanks to reduce the occupancy of V-103 and V-107; and (3) membrane filters to reduce the occupancy of DF-103 and DF-105. The additional capital investment for such a change is around $20 million. This additional investment will allow the plant's capacity to be doubled, and the new unit production cost will be around $33.6/g. The reduction in the unit production cost is rather small because the majority of the cost is associated with raw materials, consumables, and waste disposal that scale approximately linearly with production.

### Monoclonal Antibody Production

Monoclonal antibodies (MAb's) are used in diagnostic tests as well as for therapeutic purposes. World demand for currently approved MAb's is on the order of a few kilograms per year. However, new therapeutic MAb's are under development that require doses of several hundred milligrams to a gram over the course of therapy. The world demand for such products will exceed 100 kg per year.

Current production choices for MAb's are limited to three well-established systems: ascites, stirred tank bioreactors (STR), and hollow-fiber bioreactors. Alternative technologies under development include transgenic animals and genetically altered plants. Currently, stirred tank bioreactors tend to be favored for production of MAb's in kilogram quantities. They are operated under batch, fed-batch, or perfusion mode. This example analyses the production of a typical therapeutic monoclonal antibody. In the base case, approximately 6.2 kg of purified product is produced per year in 46 batches. The manufacturing cost for producing larger quantities is estimated as part of the sensitivity analysis.

*Upstream section*: The serum-free and low-protein-content media powder is dissolved in WFI in a stainless steel tank (V-101), and the solution is sterilised using a 0.1 mm dead-end polishing filter (DE-101). The concentration of media powder in the feed solution is 10 g/liter. A stirred-tank bioreactor (V-102) is used to grow the cells that express the therapeutic immunoglobulin G (IgG). The bioreactor operates in fedbatch mode. A cycle time of 156 hours (132 hours for fermentation and 24 hours for turnaround) was assumed for the bioreactor. The volume of broth generated per bioreactor batch is approximately 2,200 liters containing 220 g of product (the product titer is 100 mg/L). The total volume of the bioreactor vessel is 3,000 liters.

*Downstream section*: The generated biomass and other suspended compounds are removed using a 0.65 mm membrane diafilter (DF-101). The product recovery yield of this step is 95%. This filtration step takes 5.1 h and requires a membrane area of around 30 m2. The clarified solution is concentrated 20-fold using a 50,000 MW cut-off ultrafilter (UF-101). The recovery yield of this step is 95%. This step takes 3.6 h and requires a membrane area of 40 m2.

The bulk of the contaminant proteins are removed using a protein A affinity chromatography column (C-101). The following operating assumptions were made: (1) resin binding capacity is 15 mg of product per ml of resin; (2) the eluant is a 0.1 M solution of sodium citrate, and its volume is equal to 6 column volumes (CV's); (3) the product is recovered in 3 CV's of eluant buffer with a recovery yield of 95%, and the pH is maintained near neutral to ensure product stability; and (4) the total volume of the solutions for column equilibration, wash, and regeneration is 13 CV's. This step takes around 15.7 h and requires a resin volume of 24.5 liters.

The protein A elution buffer is exchanged with phosphate buffer using the same diafilter as in P-7. The product recovery yield of this step is 95%. The purification proceeds using a cationexchange chromatography column. The following operating assumptions were made: (1) the resin's binding capacity is 20 mg of product per ml of resin; (2) a gradient elution step is employed with a sodium chloride concentration ranging from 0.0 to 1.0 M and a volume of 6 CV's; (3) the product is recovered in 3 CV's of eluant buffer with a recovery yield of 90%; and (4) the total volume of the solutions for column equilibration, wash, and regeneration is 17 CV's. This step takes around 13 h and requires a resin volume of 15.7 liters.

Ammonium sulfate is added to a concentration of 2.0 M to increase the ionic strength of the solution and prepare it for the hydrophobic interaction chromatography (HIC) step (C-103) that follows. The following operating assumptions were made for the HIC step: (1) resin-binding capacity is 20 mg of product per ml of resin; (2) a gradient elution step is used in which the concentration of ammonium sulfate changes linearly from 2.0 M to 0.0 M; (3) the product is recovered in 2 CV's of eluant buffer with a recovery yield of 95%; and (4) the total volume of the solutions for column equilibration, wash, and regeneration is 22 CV's. This step takes around 13.5 hours and requires a resin volume of 15.7 liters.

The purified product solution is concentrated two-fold and the HIC elution buffer is exchanged with phosphate buffer using the same diafilter. Next, glycerol is added (for product stability) to a concentration of 100 g/L and the solution is sent to product formulation. The product concentration in the final solution is around 8 g/liter. The following additional assumptions were made: (1) WFI is used for the preparation of water solutions and buffers; (2) to calculate the cycle time of chromatography steps, it was assumed that loading and elution operate at a linear velocity of 100 cm/h, while equilibration, washing, and regeneration operate at a linear velocity of 200 cm/h.

The quantities are in kilograms per batch. The duration of a single batch is 162 h. The overall recovery yield of IgG (the product) is 62% (140 g of IgG is recovered out of the 220 g that is present in the fermentation broth). Note the large amount of process water and WFI utilised per batch. The majority of process water and WFI are utilised for equipment cleaning.

The plant batch time is approximately 232 h. This is the time required to go from the preparation of raw materials to the final product in a single batch. A new batch is initiated every seven days (168 h). The bioreactor, which is the time (scheduling) bottleneck, has a cycle time of 152 h (140 h for fermentation and 12 h for turnaround). While the bioreactor is preparing a new batch, the downstream equipment is being utilised to purify the product of the previous batch. The downstream section requires two shifts per day for four days a week.

Multiple bars on the same line represent reuse (sharing) of equipment by multiple procedures. White space represents idle time. On an annual basis, the plant processes 46 batches and produces 6.2 kg of purified IgG.

### Tissue Plasminogen Activator

Tissue plasminogen activator (tPA) was among the first products of biotechnology to be based on recombinant DNA technology. tPA is a medium molecular weight enzymatic protein whose primary application is in the treatment of myocardial infarction (heart attack) patients. First characterised in 1979, the protein was commercially developed by Genentech, with clinical trial quantities and purities being achieved in late 1984.

The current market demand is around 15 kg/yr of purified tPA, which generates annual revenues of around $300 million for Genentech. The typical dose of tPA is around 100 mg and this corresponds to a price per dose of around $2,000. The management of your company believes that the world demand can increase to more than

200 kg per year if the price per dose is reduced to around \$300. Before committing to the venture, your president would like you, as Director of Corporate Planning, to evaluate a technology for producing 50 kg of tPA per year using transgenic goats. Based on data from Genzyme Transgenics, you know that you can buy milk containing tPA at a concentration of 20 g/L for around \$200/g of tPA.

Based on information from the technical and patent literature, develop and evaluate a process that can recover and purify 50 kg of tPA per year from goat milk. More specifically, estimate the capital investment required and the unit production cost.

### *Indigo*

Indigo is a dye that is used by denim manufacturers. It has been traditionally produced through chemical synthesis. The chemical route, however, generates large amounts of regulated waste materials that make the process environmentally unattractive. In the late 1990's, Genencor International (a biotechnology company) commercialised a technology for producing indigo via fermentation. Based on information from the technical and patent literature, develop and evaluate a process for producing 5,000,000 kg of indigo per year via fermentation. The product must meet the quality specifications of the denim industry. More specifically, estimate the capital investment required and the unit production cost.

### *L-lysine*

L-lysine is an amino acid that is produced in large quantities (over 100,000 metric tons per year) via fermentation. It is used as an animal feed supplement mainly for poultry and pigs. Based on information from the technical and patent literature, develop and evaluate a process for producing 15,000,000 kg of L-lysine per year via fermentation. Your analysis should include estimation of capital and operating cost.

### Xanthan Gum

Xanthan gum is a water-soluble polysaccharide produced via fermentation. It is used in food products as a thickener, stabiliser and an emulsifier. Xanthan gum is also used for enhanced oil recovery. Based on information from the technical and patent literature, develop and evaluate a process for producing 10,000,000 kg of xanthan gum per year. The product should meet the specifications of the petroleum industry for enhanced oil recovery. Your analysis should include estimation of capital and operating cost. Also, perform sensitivity analysis and estimate the unit production cost for plant capacities ranging from 10 to 50 million kg of xanthan gum per year.

### Biodegradable Polymers

Because of the capacity limitations of urban landfills, biodegradable plastic packaging materials are of interest as a means to reduce the load of solid waste disposal systems. Poly-2-hydroxybutyrate (PHB), a biodegradable polyester, is such a promising material that can be produced via fermentation. Microorganisms that synthesise PHB include Gram positive and Gram negative species and cyanobacteria. Some members of the groups of Alcaligenes and Azotobacter are the most promising because they store high levels of PHB. PHB is synthesised and stored intracellularly as a possible future carbon and energy source.

High levels of polymer are obtained under nitrogen and phosphorous limitation. At optimum conditions, PHB can reach 70-80% by weight of the cell mass of the organism. Based on information from the technical and patent literature, design and evaluate a plant that produces 30,000,000 kg of PHB per year.

### *Detergent Enzymes*

Proteolytic enzymes are used in detergents to hydrolyse and remove proteinaceous stains. The commercially important proteolytic enzymes that are used in detergents are mainly produced by Bacillus subtilis and Bacillus licheniformis. These enzymes are endo-cleaving, have broad specificity, are active over a wide pH range and calcium improves their stability at high temperature or extremes of pH. Their molecular weight is around 30,000 and their isoelectric point is in the range of 8.5-9.5. The current world demand for detergent enzymes is around 10,000 metric tons (of pure enzyme) per year, corresponding to a world market of around $300 million.

The marketing department of your company believes that the world demand can increase to more than 20,000 metric tons per year if the selling price is reduced to around $15/kg of pure enzyme. Before committing to the venture, your company would like you to evaluate the cost structure of the current producers and find out if it is possible (through the use of genetic engineering and modern separation technologies) to produce such enzymes for less than $10/kg.

### *Therapeutic Proteins from Transgenic Tobacco*

Transgenic plants (e.g., corn, tobacco, etc.) have the potential to produce complex bioactive proteins at significantly lower cost than production via transgenic animals or mammalian cell cultures. The advantages of transgenic plant production are easy and efficient introduction of stable foreign genes, cost-effective biomass production ($0.02 to $0.04 per kg), no possible contamination with human disease agents, and the ability to perform complex protein processing needed for many bioactive human therapeutics.

Cost-effective biomass production makes this mode of production suitable for large volume recombinant proteins. The ability to perform complex protein processing is advantageous for production of therapeutic glycoproteins and bioactive peptides. Downstream processing costs are a major portion of the total unit production cost associated with transgenic plant production of large volume therapeutic proteins. Therefore, primary recovery requires significant volume reduction. Once volume reduction and biomass removal are achieved, chromatographic purification is required to remove plant protein impurities.

Design a purification process for use in a facility manufacturing 100 metric tons per year of a recombinant human blood protein from transgenic tobacco. Assume that the expression level is 0.5 g of product protein per kg of tobacco. Assume that biomass production and primary recovery are performed at a separate site. The feed to the purification section is 4,000 L per day (containing 200 g/L of product protein) and this material is purchased for \$0.5/g of product protein. The process should include the appropriate filtration and chromatographic steps. Assume that a combination of affinity and ion exchange chromatography provide >90% pure product.

# 3

# Enzyme Preparation

Biologically active enzymes may be extracted from any living organism. A very wide range of sources are used for commercial enzyme production from Actinoplanes to Zymomonas, from spinach to snake venom. Of the hundred or so enzymes being used industrially, over a half are from fungi and yeast and over a third are from bacteria with the remainder divided between animal (8%) and plant (4%) sources. A very much larger number of enzymes find use in chemical analysis and clinical diagnosis. Non-microbial sources provide a larger proportion of these, at the present time. Microbes are preferred to plants and animals as sources of enzymes because:

1. they are generally cheaper to produce.
2. their enzyme contents are more predictable and controllable,
3. reliable supplies of raw material of constant composition are more easily arranged, and
4. plant and animal tissues contain more potentially harmful materials than microbes, including phenolic compounds (from plants), endogenous enzyme inhibitors and proteases.

Attempts are being made to overcome some of these difficulties by the use of animal and plant cell culture.

In practice, the great majority of microbial enzymes come from a very limited number of genera, of which Aspergillus species, Bacillus species and Kluyveromyces (also called Saccharomyces) species predominate. Most of the strains used have either been employed by the food industry for many years or have been derived from such strains by mutation and selection. There are very few examples of the industrial use of enzymes having been developed for one task.

Shining examples of such developments are the production of high fructose syrup using glucose isomerase and the use of pullulanase in starch hydrolysis. Producers of industrial enzymes and their customers will share the common aims of economy, effectiveness and safety. They will wish to have high-yielding strains of microbes which make the enzyme constitutively and secrete it into their growth medium (extracellular enzymes).

If the enzyme is not produced constitutively, induction must be rapid and inexpensive. Producers will aim to use strains of microbe that are known to be generally safe. Users will pay little regard to the way in which the enzyme is produced but will insist on having preparations that have a known activity and keep that activity for extended periods, stored at room temperature or with routine refrigeration.

They will pay little attention to the purity of the enzyme preparation provided that it does not contain materials (enzymes or not) that interfere with their process. Both producers and users will wish to have the enzymes in forms that present minimal hazard to those handling them or consuming their product. The development of commercial enzymes is a specialised business which is usually undertaken by a handful of companies which have high skills in

1. screening for new and improved enzymes,
2. fermentation for enzyme production,
3. large scale enzyme purifications,
4. formulation of enzymes for sale,
5. customer liaison, and
6. dealing with the regulatory authorities.

### Novel Enzymes

If a reaction is thermodynamically possible, it is likely that an enzyme exists which is capable of catalysing it. One of the major skills of enzyme companies and suitably funded academic laboratories is the rapid and cost-effective screening of microbial cultures for enzyme activities. Natural samples, usually soil or compost material found near high concentrations of likely substrates, are used as sources of cultures. It is not unusual at international congresses of enzyme technologists to see representatives of enzyme companies collecting samples of soil to be screened later when they return to their laboratories. The first stage of the screening procedure for commercial enzymes is to screen ideas, i.e. to determine the potential commercial need for a new enzyme, to estimate the size of the market and to decide, approximately, how much potential users of the enzyme will be able to afford to pay for it.

In some cases, the determination of the potential value of an enzyme is not easy, for instance when it might be used to produce an entirely novel substance. In others, for instance when the novel enzyme would be used to improve an existing process, its potential value can be costed very accurately. In either case, a cumulative cash flow must be estimated, balancing the initial screening and investment capital costs including interest, tax liability and depreciation, against the expected long term profits. Full account must be taken of inflation, projected variation in feedstock price and source, publicity and other costs. In addition, the probability of potential market competition and changes in political or legal factors must be considered.

Usually the sensitivity of the project to changes in all of these factors must be estimated, by informed guesswork, in order to assess the risk factor involved. Financial re-appraisal must be frequently carried out during the development process to check that it still constitutes an efficient use of resources. If agreement is reached, probably after discussions with potential users, that experimental work would be commercially justifiable, the next stage involves the location of a source of the required enzyme. Laboratory work is expensive in manpower so clearly it is worthwhile using all available databases to search for mention of the enzyme in the academic and patents literature.

Cultures may then be sought from any sources so revealed. Some preparations of commercial enzymes are quite rich sources of enzymes other than the enzyme which is being offered for sale, revealing such preparations as potential inexpensive sources which are worth investigating. If these first searches are unsuccessful, it is probably necessary to screen for new microbial strains capable of performing the transformation required. This should not be a 'blind' screen: there will usually be some source of microbes that could have been exposed for countless generations to the conditions that the new enzyme should withstand or to chemicals which it is required to modify. Hence, thermophiles are sought in hot springs, osmophiles in sugar factories, organisms capable of metabolising wood preservatives in timber yards and so on.

A classic example of the detection of an enzyme by intelligent screening was the discovery of a commercially useful cyanide-degrading enzyme in the microbial pathogens of plants that contain cyanogenic glycosides. The identification of a microbial source of an enzyme is by no means the end of the story. The properties of the enzyme must be determined; i.e. temperature for optimum productivity, temperature stability profile, pH optimum and stability, kinetic constants ($K_m$, $V_{max}$), whether there is substrate or product inhibition, and the ability to withstand components of the expected feedstock other than substrate. A team of scientists, engineers and accountants must then consider the next steps.

If any of these parameters is unsatisfactory, the screen must continue until improved enzymes are located. Now that protein engineering can be seriously contemplated, an enzyme with sufficient potential value could be improved 'by design' to overcome one or two shortcomings. However, this would take a long time, at the present level of knowledge and skill, so further screening of microbes from selected sources would probably be considered more worthwhile. Once an enzyme with suitable properties has been located, various decisions must be made concerning the acceptability of the organism to the regulatory authorities, the productivity of the organism, and the way in which the enzyme is to be isolated, utilised (free or immobilised) and, if necessary, purified.

If the organism is unacceptable from a regulatory viewpoint two options exist; to eliminate that organism altogether and continue the screening operation, or to clone the enzyme into an acceptable organism. The latter approach is becoming increasingly attractive especially as cloning could also be used to increase the productivity of the fermentation process. Cloning may also be attractive when the organism originally producing the enzyme is acceptable from the health and safety point of view but whose productivity is unacceptable. However, cloning is not yet routine and invariably successful so there is still an excellent case to be made for applying conventional mutation and isolation techniques for the selection of improved strains.

It should be noted that although the technology for cloning glucose isomerase into 'routine' organisms is known, it has not yet been applied. Several of the glucose isomerase preparations used commercially consist of whole cells, or cell fragments, of the selected strains of species originally detected by screening. The use of immobilised enzymes is now familiar to industry and their advantages are well recognised so the practicality of using the new enzymes in an immobilised form will be determined early in the screening procedure. If the enzyme is produced intracellularly, the feasibility of using it without isolation and purification will be considered very seriously and strains selected for their amenability to use in this way. It should be emphasised that there will be a constant dialogue between laboratory scientists and biochemical process engineers from the earliest stages of the screening process.

Once the biochemical engineers are satisfied that their initial criteria of productivity, activity and stability can be met, the selected strain(s) of microbe will be grown in pilot plant conditions. It is only by applying the type of equipment used in full scale plants that accurate costing of processes can be achieved. Pilot studies will probably reveal imperfections, or at least areas of ignorance, that must be corrected at the laboratory scale. If this proves possible, the pilot plant will produce samples of the enzyme preparation to be used by customers who may well also be at the pilot plant stage in the development of the enzyme-utilizing process.

The enzyme pilot plant also produces samples for safety and toxicological studies provided that the pilot process is exactly similar to the full scale operation. Screening for new enzymes is expensive so that the intellectual property generated must be protected against copying by competitors. This is usually done by patenting the enzyme or its production method or, most usefully, the process in which it is to be used. Patenting will be initiated as soon as there is evidence that an innovative discovery has been made.

The effects of changing feedstocks must be considered in relation to downstream processing. If such variability is likely to significantly reduce the efficiency of the standard methodology, it may be economical to use a more expensive defined medium of easily reproducible composition. Clearly defined media are usually out of the question for large scale use on cost grounds but may be perfectly acceptable when enzymes are to be produced for high value uses, such as analysis or medical therapy where very pure preparations are essential.

Less-defined complex media are composed of ingredients selected on the basis of cost and availability as well as composition. Waste materials and by-products from the food and agricultural industries are often major ingredients. Thus molasses, corn steep liquor, distillers solubles and wheat bran are important components of fermentation media providing carbohydrate, minerals, nitrogen and some vitamins. Extra carbohydrate is usually supplied as starch, sometimes refined but often simply as ground cereal grains. Soybean meal and ammonium salts are frequently used sources of additional nitrogen. Most of these materials will vary in quality and composition from batch to batch causing changes in enzyme productivity.

**Enzymes Preparation**

Enzyme may be purified several hundred-fold but the yield of the enzyme may be very poor, frequently below 10% of the activity of the original material. In contrast, industrial enzymes will be purified as little as possible, only other enzymes and material likely to interfere with the process which the enzyme is to catalyse, will be removed. Unnecessary purification will be avoided as each additional stage is costly in terms of equipment, manpower and loss of enzyme activity.

As a result, some commercial enzyme preparations consist essentially of concentrated fermentation broth, plus additives to stabilise the enzyme's activity. The content of the required enzyme should be as high as possible (e.g. 10% w/w of the protein) in order to ease the downstream processing task. This may be achieved by developing the fermentation conditions or, often more dramatically, by genetic engineering.

It may well be economically viable to spend some time cloning extra copies of the required gene together with a powerful promoter back into the producing organism in order to get 'over-producers'. It is important that the maximum activity is retained during the preparation of enzymes. Enzyme inactivation can be caused by heat, proteolysis, sub-optimal pH, oxidation, denaturants, irreversible inhibitors and loss of cofactors or coenzymes.

Of these heat inactivation, which together with associated pH effects, is probably the most significant. It is likely to occur during enzyme extraction and purification if insufficient cooling is available, but the problem is less when preparing thermophilic enzymes. Proteolysis is most likely to occur in the early stages of extraction and purification when the proteases responsible for protein turnover in living cells are still present. It is also the major reason for enzyme inactivation by microbial contamination. In their native conformations, enzymes have highly structured domains which are resistant to attack by proteases because many of the peptide bonds are mechanically inaccessible and because many proteases are highly specific.

The chances of a susceptible peptide bond in a structured domain being available for protease attack are low. Single 'nicks' by proteases in these circumstances may have little immediate effect on protein conformation and, therefore, activity. The effect, however, may severely reduce the conformational stability of the enzyme to heat or pH variation so greatly reducing its operational stability. If the domain is unfolded under these changed conditions, the whole polypeptide chain may be available for proteolysis and the same, specific, protease may destroy it.

Clearly the best way of preventing proteolysis is to rapidly remove, or inhibit, protease activity. Before this can be achieved it is important to keep enzyme preparations cold to maintain their native conformation and slow any protease action that may occur. Some intracellular enzymes are used commercially without isolation and purification but the majority of commercial enzymes are either produced extracellularly by the microbe or plant or must be released from the cells into solution and further processed (Figure 1).

Solid/liquid separation is generally required for the initial separation of cell mass, the removal of cell debris after cell breakage and the collection of precipitates. This can be achieved by filtration, centrifugation or aqueous biphasic partition. In general, filtration or aqueous biphasic systems are used to remove unwanted cells or cell debris whereas centrifugation is the preferred method for the collection of required solid material.

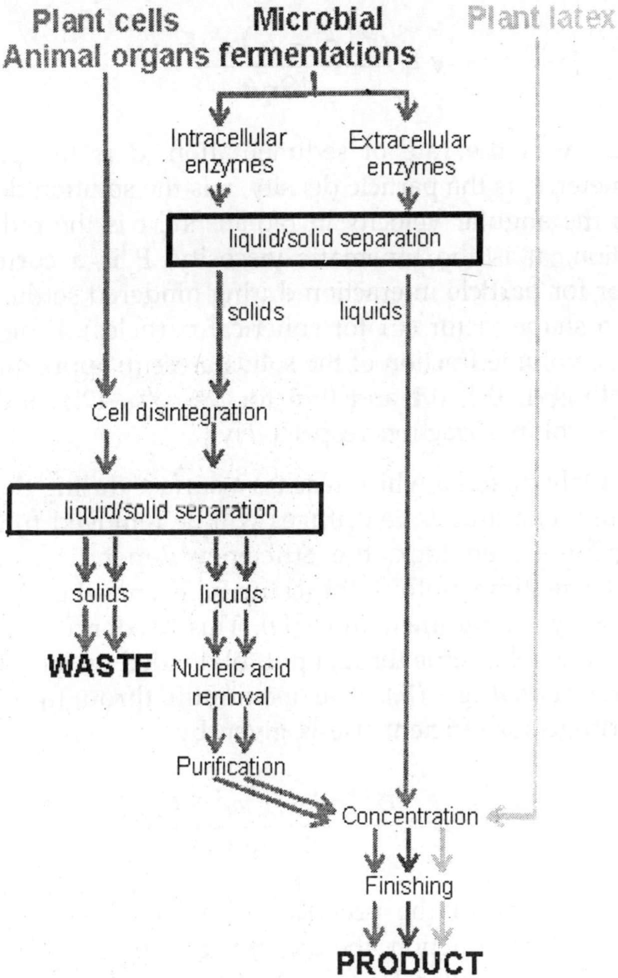


*Figure 1. Flow diagram for the preparation of enzymes.*

#### *Centrifugation*

Centrifugation separates on the basis of the particle size

and density difference between the liquid and solid phases. Sedimentation of material in a centrifugal field may be described by

$$v = \frac{d^2(\rho_s - \rho_l)\omega^2 r F_s}{18\eta\theta} \tag{1}$$

where v is the rate of sedimentation, d is the particle diameter, $r_s$ is the particle density, $r_l$ is the solution density, w is the angular velocity in radians $s^{-1}$, r is the radius of rotation, η is the kinematic viscosity, $F_s$ is a correction factor for particle interaction during hindered settling and q is a shape factor (=1 for spherical particles). $F_s$ depends on the volume fraction of the solids present; approximately equalling 1, 0.5, 0.1 and 0.05 for 1%, 3%, 12% and 20% solids volume fraction respectively.

Only material which reaches a surface during the flow through continuous centrifuges will be removed from the centrifuge feedstock, the efficiency depending on the residence time within the centrifuge and the distance necessary for sedimentation (D). This residence time will equal the volumetric throughput (ϕ) divided by the volume of the centrifuge (V). The maximum throughput of a centrifuge for efficient use is given by

$$\phi = \frac{d^2(\rho_s - \rho_l)\omega^2 r V F_s}{18\eta\theta D} \tag{2}$$

The efficiency of the process is seen to depend on the solids volume fraction, the effective clarifying surface (V/D) and the acceleration factor ($\omega^2 r/g$, where g is the gravitational constant, 981 cm $s^{-2}$; a rotor of radius 25 cm spinning at 1 rev $s^{-1}$ has an acceleration factor of approximately 1 G). Low acceleration factors of about 1 500 g may be used for harvesting cells whereas much higher acceleration factors are needed to collect enzyme efficiently. The product of these factors ($\omega^2rV/gD$) is called the sigma factor (S) and is used to compare centrifuges and to assist scale-up.

Laboratory centrifuges using tubes in swing-out or angle head rotors have high angular velocity (w) and radius of rotation (r) but small capacity (V) and substantial sedimentation distance (D). This type of design cannot be scaled-up safely, primarily because the mechanical stress on the centrifuge head increases with the square of the radius, which must increase with increasing capacity.

For large-scale use, continuous centrifuges of various types are employed (Figure 2). These allow the continuous addition of feedstock, the continuous removal of supernatant and the discontinuous, semicontinuous or continuous removal of solids. Where discontinuous or semicontinuous removal of precipitate occurs, the precipitate is flushed out by automatic discharge systems which cause its dilution with water or medium and may be a problem if the precipitate is required for further treatment. Centrifugation is the generally preferred method for the collection of enzyme-containing solids as it does not present a great hazard to most enzymes so long as foam production, with consequent enzymic inactivation, is minimised.

Small particles of cell debris and precipitated protein may be sedimented using tubular bowl centrifuges, of which Sharples centrifuges are the best known. These semi-continuous centrifuges are long and thin enabling rapid acceleration and deceleration, minimising the down-time required for the removal of the sedimented solids. Here the radius and effective liquid thickness are both small allowing a high angular velocity and hence high centrifugal force; small models can be used at acceleration factors up to 50,000 g, accumulating 0.1 Kg of wet deposit whereas large models, designed to accumulate up to 5 Kg of deposit, are restricted to 16,000 g. The capacities of these centrifuges are only moderate.

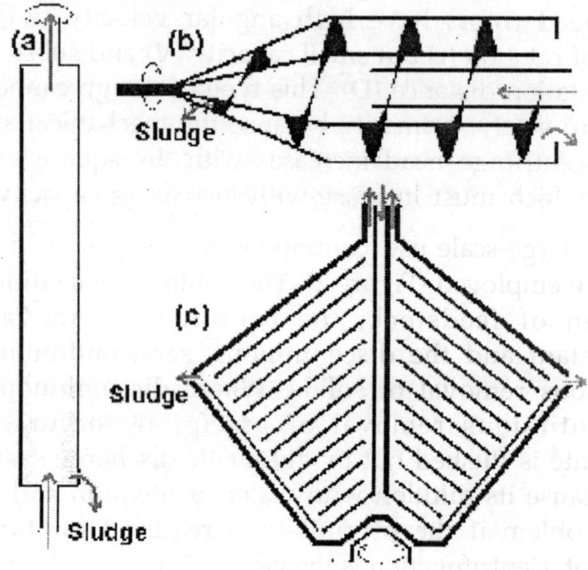


*Figure 2. Basic designs of industrial centrifuges, showing the flow of material within the bowls. (a) Tubular bowl centrifuge. (b) Continuous scroll centrifuge. (c) Continuous multichamber disc-stack centrifuge.*

Multichamber disc-stack centrifuges, originally designed (by Westfalia and Alpha-Laval) for cream separation, contain multiple coned discs in a stack which are spun and on which the precipitate collects. They may be operated either semi-continuously or, by using a centripetal pressurising pump within the centrifuge bowl which forces the sludge out through a valve, continuously. The capacity and radius of such devices are large and the thickness of liquid is very small, due to the large effective surface area.

The angular velocity, however, is restricted giving a maximum acceleration factor of about 8,000 g. A different design which is rather similar in principle is the solid bowl scroll centrifuge in which an Archimedes' screw collects the precipitate so that fluid and solids leave at opposite ends of the apparatus. These can only be used at low acceleration (about 3,000 g) so they are suitable only for the collection of comparatively large particles.

Although many types of centrifuge are available, the efficient precipitation of small particles of cell debris can be difficult, sometimes near-impossible. Clearly from Equation 2, the efficiency of centrifugation can be improved if the particle diameter (d) is increased. This can be done either by coagulating or flocculating particles.

Coagulation is caused by the removal of electrostatic charges (e.g. by pH change) and allowing particles to adhere to each other. Flocculation is achieved by adding small amounts of high-molecular-weight charged materials which bridge oppositely-charged particles to produce a loose aggregate which may be readily removed by centrifugation or filtration.

Flocculation and coagulation are cheap and effective aids to precipitating or otherwise harvesting whole cells, cell debris or soluble proteins but, of course, it is essential that the agents used must not inhibit the target enzymes. It is important to note that the choice of flocculant is determined by the pH and ionic strength of the solution and the nature of the particles. Most flocculants have very definite optimum concentrations above which further addition may be counter-effective.

Some flocculants can be rapidly ruined by shear. A comparatively recent introduction designed for the removal of cell debris is a moderately hydrophobic product in which cellulose is lightly derivatised with diethylaminoethyl functional groups. This material (Whatman CDR; cell debris remover) is inexpensive (essential as it is not reusable), binds to unwanted negatively charged cell constituents, acts as a filter aid and may be incinerated to dispose of hazardous wastes.

### Filtration

Filtration separates simply on the basis of particle size. Its efficiency is limited by the shape and compressibility of the particles, the viscosity of the liquid phase and the maximum allowable pressures. Large-scale simple filtration employs filter cloths and filter aids in a plate and frame press configuration, in rotary vacuum filters or centrifugal filters (Figure 3). The volumetric throughput of a filter is proportional to the pressure (P) and filter area ($A_F$) and inversely proportional to the filter cake thickness ($D_F$) and the dynamic viscosity

$$\phi = \frac{kPA_F}{\eta D_F} \tag{3}$$

where k is a proportionality constant dependent on the size and nature of the particles. For very small particles k depends on the fourth power of their diameter. Filtration of particles that are easily compressed leads to filter blockage and the failure of Equation 3 to describe the system. Under these circumstances a filter aid, such as celite, is mixed with the feedstock to improve the mechanical stability of the filter cake.

Filter aids are generally used only where the liquid phase is required as they cause substantial problems in the recovery of solids. They also may cause loss of enzyme activity from the solution due to physical hold-up in the filter cake. It is often difficult for a process development manager to decide whether to attempt to recover enzyme trapped in this way.

Problems associated with the build-up of the filter cake may also be avoided by high tangential flow of the feedstock across the surface of the filter, a process known as crossflow microfiltration (Figure 4). This method dispenses with filter aids and uses special symmetric microporous membrane assemblies capable of retaining particles down to 0.1 - 1 mm diameter.

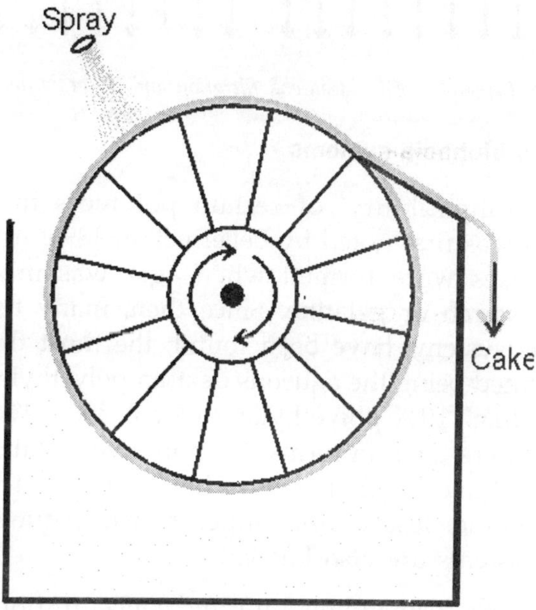


*Figure 3. The basic design of the rotary vacuum filter*

A simple and familiar filtration apparatus is the perforate bowl centrifuge or basket centrifuge, in effect a spin drier. Cell debris is collected on a cloth with, or without, filter aid and can be skimmed off when necessary using a suitable blade. Such centrifugal filters have a large radius and effective liquid depth, allowing high volumes. However, safety decrees that the angular velocity must be low and so only large particles (e.g. plant material) can be removed satisfactorily.

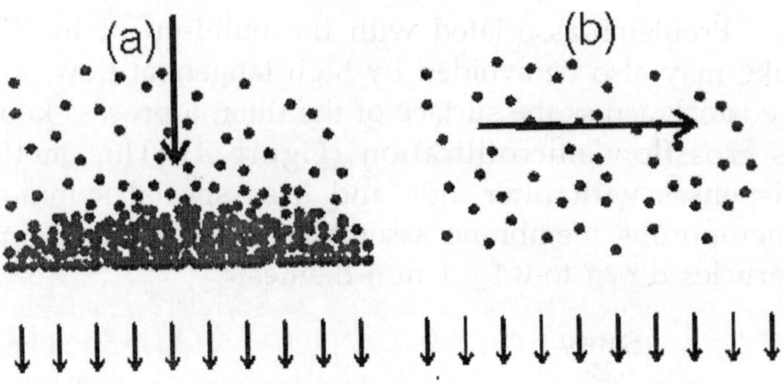


*Figure 4. Principles of (a) dead-end filtration and (b) cross-flow filtration*

### Aqueous biphasic systems

The 'incompatibility' of certain polymers in aqueous solution was first noted by Beijerinck in 1896. In this case two phases were formed when agar was mixed with soluble starch or gelatine. Since then, many two phase aqueous systems have been found; the most thoroughly investigated being the aqueous dextran-polyethylene glycol system (e.g. 10% polyethylene glycol 4000/2% dextran T500), where dextran forms the more hydrophilic, denser, lower phase and polyethylene glycol the more hydrophobic, less dense, upper phase. Aqueous three phase systems are also known.

Phases form when limiting concentrations of the polymers are exceeded. Both phases contain mainly water and are enriched in one of the polymers. The limiting concentrations depend on the type and molecular weight of the polymers and on the pH, ionic strength and temperature of the solution. Some polymers form the upper hydrophobic phase in the presence of fairly concentrated solutions of phosphates or sulphates (e.g. 10% polyethylene glycol 4000/12.5% potassium phosphate buffer). A drawback to the useful dextran/polyethylene glycol system is the high cost of the purified dextran used. This has been alleviated by the use of crud unfractionated dextran preparations, much cheaper hydroxypropyl starch derivatives and salt-containing biphasic systems.

Aqueous biphasic systems are of considerable value to biotechnology. They provide the opportunity for the rapid separation of biological materials with little probability of denaturation. The interfacial tension between the phases is very low (i.e. about 400-fold less than that between water and an immiscible organic solvent), allowing small droplet size, large interfacial areas, efficient mixing under very gentle stirring and rapid partition. The polymers have a stabilising influence on most proteins. A great variety of separations have been achieved, by far the most important being the separation of enzymes from broken crude cell material.

Separation may be achieved in a few minutes, minimising the harmful action of endogenous proteases. The systems have also been used successfully for the separation of different types of cell membranes and organelles, the purification of enzymes and for extractive bioconversions. Continuous liquid two-phase separation is easier than continuous solid/liquid separation using equipment familiar from immiscible solvent systems, for example disc-stack centrifuges and counter-current separators. Such systems are readily amenable to scale-up and may be employed in continuous enzyme extraction processes involving some recycling of the phases.

Cells, cell debris proteins and other material distribute themselves between the two phases in a manner described by the partition coefficient (P) defined as:

$$P = \frac{C_t}{C_b} \qquad (4)$$

where $C_t$ and $C_b$ represent the concentrations in the top and bottom phases respectively. The yield and efficiency of the separation is determined by the relative amounts of material in the two phases and therefore depends on the volume ratio ($V_t/V_b$). The partition coefficient is exponentially related to the surface area (and hence molecular weight) and surface charge of the particles in addition to the difference in the electrical potential and hydrophobicity of the phases. It is not generally very sensitive to temperature changes. This means that proteins and larger particles are normally partitioned into one phase whereas smaller molecules are distributed more evenly between phases. A partition coefficient of greater than 3 is required if usable yields are to be achieved by a single extraction process.

Typical partition coefficients for proteins are 0.01-100 whereas the partition coefficients for cells and cell debris are effectively zero. The influence of pH and salts on protein partition is complex, particularly when phosphate buffers are present. A given protein distributes differently between the phases at different pH's and ionic strength but the presence of phosphate ions affect the partition coefficient in an anomalous fashion because these ions distribute themselves unequally resulting in electrostatic potential (and pH) differences. This means that systems may be 'tuned' to enrich an enzyme in one phase, ideally the upper phase with cell debris and unwanted enzymes in the lower phase.

An enzyme may be extracted from the upper (polyethylene glycol) phase by the addition of salts or further polymer, generating a new biphasic system. This stage may be used to further purify the enzyme. A powerful modification of this technique is to combine phase partitioning and affinity partitioning. Affinity ligands (e.g. triazine dyes) may be coupled to either polymer in an aqueous biphasic system and thus greatly increase the specificity of the extraction.

**Breakage of Cell**

Various intracellular enzymes are used in significant quantities and must be released from cells and purified. The amount of energy that must be put into the breakage of cells depends very much on the type of organism and to some extent on the physiology of the organism. Some types of cell are broken readily by gentle treatment such as osmotic shock (e.g. animal cells and some gram-negative bacteria such as Azotobacter species), whilst others are highly resistant to breakage. These include yeasts, green algae, fungal mycelia and some gram-positive bacteria which have cell wall and membrane structures capable of resisting internal osmotic pressures of around 20 atmospheres (2 MPa) and therefore have the strength, weight for weight, of reinforced concrete.

Consequently a variety of cell disruption techniques have been developed involving solid or liquid shear or cell lysis. The rate of protein released by mechanical cell disruption is usually found to be proportional to the amount of releasable protein.

$$\frac{dP}{dt} = -kP \tag{5}$$

where P represents the protein content remaining associated with the cells, t is the time and k is a release constant dependent on the system. Integrating from P = $P_m$ (maximum possible protein releasable) at time zero to P = $P_t$ at time t gives

$$\int_{P_m}^{P_t} \frac{dP}{P} = \int_{0}^{t} -k\,dt \tag{6}$$

$$Ln\left(\frac{P_m}{P_t}\right) = kt \tag{7}$$

As the protein released from the cells ($P_r$) is given by

$$P_r = P_m - P_t \tag{8}$$

the following equation for cell breakage is obtained

$$Ln\left(\frac{P_m}{P_m - P_r}\right) = kt \tag{9}$$

It is most important in choosing cell disruption strategies to avoid damaging the enzymes. The particular hazards to enzyme activity relevant to cell breakage are summarised in Table 1. The most significant of these, in general, are heating and shear.

*Table 1. Hazards likely to damage enzymes during cell disruption.*

| | |
|---|---|
| Heat | All mechanical methods require a large input of energy, generating heat. Cooling is essential for most enzymes. The presence of substrates, substrate analogues or polyols may also help stabilise the enzyme. |
| Shear | Shear forces are needed to disrupt cells and may damage enzymes, particularly in the presence of heavy metal ions and/or an air interface. |
| Proteases | Disruption of cells will inevitably release degradative enzymes which may cause serious loss of enzyme activity. Such action may be minimised by increased speed of processing with as much cooling as possible. This may be improved by the presence of an excess of alternative substrates or inhibitors in the extraction medium. |
| pH | Buffered solutions may be necessary. The presence of substrates, substrate analogues or polyols may also help stabilise the enzyme. |
| Chemical | Some enzymes may suffer conformational changes in the presence of detergent and/or solvents. Polyphenolics derived from plants are potent inhibitors of enzymes. This problem may be overcome by the use of adsorbents. |

| | |
|---|---|
| Oxidation | Reducing agents (e.g. ascorbic acid, mercaptoethanol and dithiothreitol) may be necessary. |
| Foaming | The gas-liquid phase interfaces present in foams may disrupt enzyme conformation. |
| Heavy-metal | Heavy metal ions (e.g. iron, copper and nickel) may toxicity be introduced by leaching from the homogenisation apparatus. Enzymes may be protected from irreversible inactivation by the use of chelating reagents, such as EDTA. |

Media for enzyme extraction will be selected on the basis of cost-effectiveness so will include as few components as possible. Media will usually be buffered at a pH value which has been determined to give the maximum stability of the enzyme to be extracted. Other components will combat other hazards to the enzyme, primarily factors causing denaturation (Table 1).

## Cell Disruption

The treatment of microbial cells in suspension with inaudible ultrasound (greater than about 18 kHz) results in their inactivation and disruption. Ultrasonication utilises the rapid sinusoidal movement of a probe within the liquid.

Ultrasonication is characterised by high frequency (18 kHz - 1 MHz), small displacements (less than about 50 mm), moderate velocities (a few m $s^{-1}$), steep transverse velocity gradients (up to 4,000 $s^{-1}$) and very high acceleration (up to about 80,000 g). It produces cavitation phenomena when acoustic power inputs are sufficiently high to allow the multiple production of microbubbles at nucleation sites in the fluid.

The bubbles grow during the rarefying phase of the sound wave, then are collapsed during the compression phase. On collapse, a violent shock wave passes through the medium. The whole process of gas bubble nucleation, growth and collapse due to the action of intense sound waves is called cavitation. The collapse of the bubbles converts sonic energy into mechanical energy in the form of shock waves equivalent to several thousand atmospheres (300 MPa) pressure. This energy imparts motions to parts of cells which disintegrate when their kinetic energy content exceeds the wall strength.

An additional factor which increases cell breakage is the microstreaming (very high velocity gradients causing shear stress) which occur near radially vibrating bubbles of gas caused by the ultrasound. Much of the energy absorbed by cell suspensions is converted to heat so effective cooling is essential. The amount of protein released by sonication has been shown to follow Equation 9. The constant (k) is independent of cell concentrations up to high levels and approximately proportional to the input acoustic power above the threshold power necessary for cavitation.

Disintegration is independent of the sonication frequency except insofar as the cavitation threshold frequency depends on the frequency. Equipment for the large-scale continuous use of ultrasonics has been available for many years and is widely used by the chemical industry but has not yet found extensive use in enzyme production. Reasons for this may be the conformational lability of some (perhaps most) enzymes to sonication and the damage that they may realise though oxidation by the free radicals, singlet oxygen and hydrogen peroxide that may be concomitantly produced. Use of radical scavengers (e.g. $N_2O$) have been shown to reduce this inactivation. As with most cell breakage methods, very fine cell debris particles may be produced which can hinder further processing. Sonication remains, however, a popular, useful and simple small-scale method for cell disruption.

### High Pressure Homogenisers

Various types of high pressure homogeniser are available for use in the food and chemicals industries but the design which has been very extensively used for cell disruption is the Manton-Gaulin APV type homogeniser. This consists of a positive displacement pump which draws cell suspension (about 12% w/v) through a check valve into the pump cylinder and forces it, at high pressures of up to 150 MPa (10 tons per square inch) and flow rates of up to 10,000 L $hr^{-1}$, through an adjustable discharge valve which has a restricted orifice (Figure 5). Cells are subjected to impact, shear and a severe pressure drop across the valve but the precise mechanism of cell disruption is not clear.

The main disruptive factor is the pressure applied and consequent pressure drop across the valve. This causes the impact and shear stress which are proportional to the operating pressure.

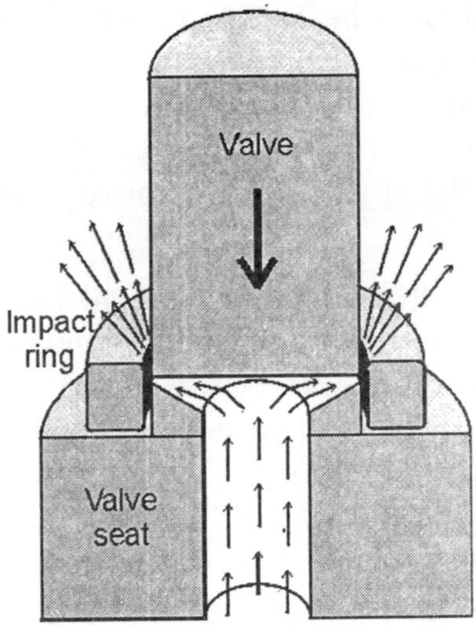


*Figure 5. A cross-section through the Manton-Gaulin homogeniser valve, showing the flow of material.*

As narrow orifices which are vulnerable to blockage are key parts of this type of homogeniser, it is unsuitable for the disruption of mycelial organisms but has been used extensively for the disruption of unicellular organisms. The release of proteins can be described by Equation 9 but normally a similar relationship is used where the time variable is replaced by the number of passes (N) through the homogeniser.

$$Ln\left(\frac{P_m}{P_m - P_r}\right) = kN \qquad (10)$$

In the commonly-used operating range with pressures below about 75 MPa, the release constant (k) has been found to be proportional to the pressure raised to an exponent dependent on the organism and its growth history (e.g. $k=k'P^{2.9}$ in Saccharomyces cerevesiae and $k=k'P^{2.2}$ in Escherichia coli, where P represents the operating pressure and k' is a rate constant). Different growth media may be selected to give rise to cells of different cell wall strength. Clearly, the higher the operating pressure, the more efficient is the disruption process. The protein release rate constant (k) is temperature dependent, disruption being more rapid at higher temperatures.

In practice, this advantage cannot be used since the temperature rise due to adiabatic compression is very significant so samples must be pre-cooled and cooled again between multiple passes. At an operating pressure of 50 MPa, the temperature rise each pass is about 12 deg. C. In addition to the fragility of the cells, the location of an enzyme within the cells can influence the conditions of use of an homogeniser. Unbound intracellular enzymes may be released by a single pass whereas membrane bound enzymes require several passes for reasonable yields to be obtained.

Multiple passes are undesirable because, of course, they decrease the throughput productivity rate and because the further passage of already broken cells results in fine debris which is excessively difficult to remove further downstream. Consequently, homogenisers will be used at the highest pressures compatible with the reliability and safety of the equipment and the temperature stability of the enzyme(s) released. High pressure homogenisers are acceptably good for the disruption of unicellular organisms provided the enzymes needed are not heat labile. The shear forces produced are not capable of damaging enzymes free in solution. The valve unit is prone to erosion and must be precision made and well maintained.

### Bead Mills

When cell suspensions are agitated in the presence of small steel or glass beads (usually 0.2 -.1.0 mm diameter) they are broken by the high liquid shear gradients and collision with the beads. The rate and effectiveness of enzyme release can be modified by changing the rates of agitation and the size of the beads, as well as the dimensions of the equipment. Any type of biomass, filamentous or unicellular, may be disrupted by bead milling but, in general, the larger sized cells will be broken more readily than small bacteria. For the same volume of beads, a large number of small beads will be more effective than a relatively small number of larger beads because of the increased likelihood of collisions between beads and cells.

Bead mills are available in various sizes and configurations from the Mickle shaker which has a maximum volume of about 40 ml to continuous process equipment capable of handling up to 200 Kg wet yeast or 20 Kg wet bacteria each hour. The bead mills that have been studied in most detail are the Dyno-Mill and the Netsch-Molinex agitator, both of which consist of a cylindrical vessel containing a motor-driven central shaft equipped with impellers of different types. Both can be operated continuously, being equipped with devices which retain the beads within the milling chamber. Glass Ballotini or stainless steel balls are used, the size range being selected for most effective release of the enzyme required. Thus 1 mm diameter beads are satisfactory for the rapid release of periplasmic enzymes from yeast but 0.25 mm diameter beads must be used, for a longer period, to release membrane-bound enzymes from bacteria.

The kinetics of protein release from bead mills follows the relationship given by Equation 9 with respect to the time (t) that a particle spends in the mill. Unfortunately, however well designed these mills are, when continuously operated there will be a significant amount of backmixing which reduces the efficiency of the protein released with respect to the average residence time (t). This is more noticeable at low flow rates (high average residence times) and when the proportion of protein released is high. It may be counteracted by designing the bead mill to encourage plug flow characteristics. Under these circumstances the relationship can be shown to be

$$Ln\left(\frac{P_m}{P_m - P_r}\right) = \frac{Ln(1 + ik\tau)}{i} \qquad (11)$$

where i represents the degree of backmixing (i.e. i = 0 under ideal plug flow conditions and i = 1 for ideal complete backmixing). Equation 11 reduces to give the simplified relationship of Equation 9 at low (near zero) values of i. In addition to bead size, the protein release rate constant (k) is a function of temperature, bead loading, impeller rotational speed and cell loading.

Impeller speeds can be increased with advantage until bead breakage becomes significant but heat generation will also increase. At a constant impeller speed, the efficiency of the equipment declines with throughput as the degree of backmixing increases. There will be an optimum impeller tip speed at which the increases in disruption are balanced by increases in backmixing.

In general, increased bead loading increases the rate of protein release but also increases the production of heat and the power consumption. Heat production is the major problem in the use of bead mills for enzyme release, particularly on a large (e.g. 20 litres) scale. Smaller vessels may be cooled adequately through cooling jackets around the bead chamber but larger mills require cooling through the agitator shaft and impellers. However, if cooling is effective there is little damage to the enzymes released.

### Freeze-presses

The Hughes press and the 'X' press enable frozen cell pastes to be forced under high pressure (150 - 230 MPa, 10 - 15 tons per square inch) through narrow orifices, the disruption being produced by phase and volume changes and by solid shear due to the ice crystals. The Hughes press can only be used discontinuously on a small scale. The 'X' press may be used semi-continuously and is amenable to scale-up but, although the method allows the breakage of even the most robust organisms and the efficient recovery of heat-sensitive enzymes, freeze-pressing is not used on a large scale for releasing enzymes from cells.

### Lytic Methods

The breakage of cells using non-mechanical methods is attractive in that it offers the prospects of releasing enzymes under conditions that are gentle, do not subject the enzyme to heat or shear, may be very cheap, and are quiet to the user. The methods that are available include osmotic shock, freezing followed by thawing, cold shock, desiccation, enzymic lysis and chemical lysis. Each method has its drawbacks but may be particularly useful under certain specific circumstances. Certain types of cell can be caused to lyse by osmotic shock. This would be a cheap, gentle and convenient method of releasing enzymes but has not apparently been used on a large scale. Some types of cell may be caused to autolyse, in particular yeasts and Bacillus species.

Yeast invertase preparations employed in the industrial manufacture of invert sugars are produced in this manner. Autolysis is a slow process compared with mechanical methods, and microbial contamination is a potential hazard, but it can be used on a very large scale if necessary. Where applicable, dessication may be very useful in the preparation of enzymes on a large scale. The rate of drying is very important in these cases, slow methods being preferred to rapid ones like lyophilisation. Enzymic lysis using added enzymes has been used widely on the laboratory scale but is less popular for industrial purposes.

Lysozyme, from hen egg-white, is the only lytic enzyme available on a commercial scale. It has often used to lyse Gram positive bacteria in an hour at about 50,000 U $Kg^{-1}$ (dry weight). The chief objection to its use on a large scale is its cost. Where costs are reduced by the use of the relatively inexpensive, lysozyme-rich, dried egg white, a major separation problem may be introduced. Yeast-lytic enzymes from Cytophaga species have been studied in some detail and other lytic enzymes are under development.

If significant markets for lytic enzymes are identified, the scale of their production will increase and their cost is likely to decrease. Lysis by acid, alkali, surfactants and solvents can be effective in releasing enzymes, provided that the enzymes are sufficiently robust. Detergents, such as Triton X-100, used alone or in combination with certain chaotropic agents, such as guanidine HCl, are effective in releasing membrane-bound enzymes. However, such materials are costly and may be difficult to remove from the final product.

# 4

# Protein Purification

Protein purification varies from simple one-step precipitation procedures to large scale validated production processes. Often more than one purification step is necessary to reach the desired purity. The key to successful and efficient protein purification is to select the most appropriate techniques, optimise their performance to suit the requirements and combine them in a logical way to maximise yield and minimise the number of steps required.

Most purification schemes involve some form of chromatography. As a result chromatography has become an essential tool in every laboratory where protein purification is needed. Different chromatography techniques with different selectivities can form powerful combinations for the purification of any biomolecule The development of recombinant DNA techniques has revolutionised the production of proteins in large quantities.

Recombinant proteins are often produced in forms which facilitate their subsequent chromatographic purification. However, this has not removed all challenges. Host contaminants are still present and problems related to solubility, structural integrity and biological activity can still exist.

Most purification protocols require more than one step to achieve the desired level of product purity. This includes any conditioning steps necessary to transfer the product from one technique into conditions suitable to perform the next technique. Each step in the process will cause some loss of product. For example, if a yield of 80% in each step is assumed, this will be reduced to only 20% overall yield after 8 processing steps as shown in Figure 1.

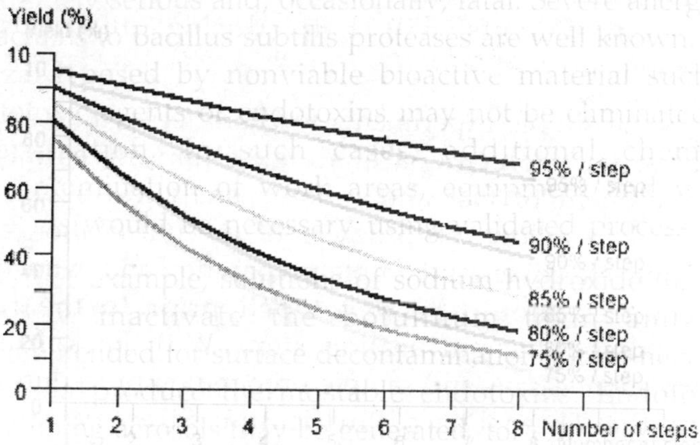


*Figure 1. Yields from multi-step purifications.*

Consequently, to reach the targets for yield and purity with the minimum number of steps and the simplest possible design, it is not efficient to add one step to another until purity requirements have been fulfilled. Occasionally when a sample is readily available purity can be achieved by simply adding or repeating steps. However, experience shows that, even for the most challenging applications, high purity and yield can be achieved efficiently in fewer than four well-chosen and optimised purification steps. Techniques should be organised in a logical sequence to avoid the need for conditioning steps and the chromatographic techniques selected appropriately to use as few purification steps as possible.

The need to obtain a protein, efficiently, economically and in sufficient purity and quantity, applies to every purification. It is important to set objectives for purity, quantity and maintenance of biological activity and to define the economical and time framework for the work. All information concerning properties of the target protein and contaminants will help during purification development.

Some simple experiments to characterise the sample and target molecule are an excellent investment. Development of fast and reliable analytical assays is essential to follow the progress of the purification and assess its effectiveness. Sample preparation and extraction procedures should be developed prior to the first chromatographic purification step. With background information, assays and sample preparation procedures in place the Three Phase Purification Strategy can be considered.

## Phase Purification Strategy

The purification has three phases Capture, Intermediate Purification and Polishing. In the Three Phase Strategy specific objectives are assigned to each step within the process:

— In the capture phase the objectives are to isolate, concentrate and stabilise the target product.

— During the intermediate purification phase the objective is to remove most of the bulk impurities such as other proteins and nucleic acids, endotoxins and viruses.

— In the polishing phase the objective is to achieve high purity by removing any remaining trace impurities or closely related substances.

The selection and optimum combination of purification techniques for Capture, Intermediate Purification and Polishing is crucial to ensure fast method development, a shorter time to pure product and good economy. The final purification process should ideally consist of sample preparation, including extraction and clarification when required, followed by three major purification steps, as shown in Figure 2. The number of steps used will always depend upon the purity requirements and intended use for the protein.

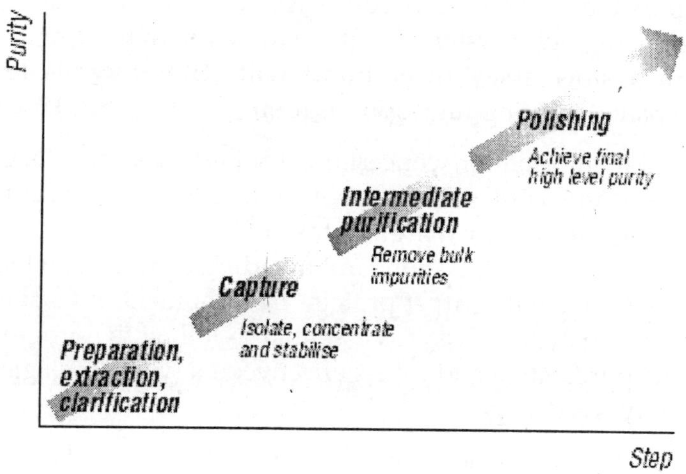


*Figure 2. Preparation and the Three Phase Purification Strategy*

## Preparation

The need to obtain a protein, efficiently, economically and in sufficient purity and quantity, applies to any purification, from preparation of an enriched protein extract for biochemical characterisation to large scale production of a therapeutic recombinant protein. It is important to set objectives for purity and quantity, maintenance of biological activity and economy in terms of money and time. Purity requirements must take into consideration the nature of the source material, the intended use of the final product and any special safety issues.

For example, it is important to differentiate between contaminants which must be removed and those which can be tolerated. Other factors can also influence the prioritisation of objectives. High yields are usually a key objective, but may be less crucial in cases where a sample is readily available or product is required only in small quantities. Extensive method development may be impossible without resources such as an ÄKTA™design chromatography system. Similarly, time pressure combined with a slow assay turnaround will steer towards less extensive scouting and optimisation.

All information concerning properties of the target protein and contaminants will help during purification development, allowing faster and easier technique selection and optimisation, and avoiding conditions which may inactivate the target protein. Development of fast and reliable analytical assays is essential to follow the progress of the purification and assess effectiveness (yield, biological activity, recovery).

*Properties of Target Protein and Critical Impurities*: All information concerning the target protein and contaminant properties will help to guide the choice of separation techniques and experimental conditions for purification. Database information for the target, or related proteins, may give size, isoelectric point (pI) and hydrophobicity or solubility data. Native one and two dimensional PAGE can indicate sample complexity and the properties of the target protein and major contaminants. Particularly important is a knowledge of the stability window of the protein so that irreversible inactivation is avoided. It is advisable to check the target protein stability window for at least pH and ionic strength. Table 1 shows how different target protein properties can affect a purification strategy.

*Goal*: To determine a 'stability window' for the target protein for easier selection and optimisation of techniques and to avoid protein inactivation during purification.

*Develop Analytical Assays*: To progress efficiently during method development the effectiveness of each step should be assessed. The laboratory should have access to the following assays:

— A rapid, reliable assay for the target protein

— Purity determination

— Total protein determination

— Assays for impurities which must be removed

The importance of a reliable assay for the target protein cannot be over-emphasised. When testing chromatographic fractions ensure that the buffers used for separation do not interfere with the assay. Purity of the target protein is most often estimated by SDS-PAGE, capillary electrophoresis, reversed phase chromatography or mass spectrometry.

Lowry or Bradford assays are used most frequently to determine the total protein. The Bradford assay is particularly suited to samples where there is a high lipid content which may interfere with the Lowry assay. For large scale protein purification the need to assay for target proteins and critical impurities is often essential. In practice, when a protein is purified for research purposes, it is too time consuming to identify and set up specific assays for harmful contaminants. A practical approach is to purify the protein to a certain level, and then perform

SDS-PAGE after a storage period to check for protease cleavage. Suitable control experiments, included within assays for bio-activity, will help to indicate if impurities are interfering with results.

*Goal*: To follow the progress of a purification, to assess effectiveness (yield, biological activity, recovery) and to help during optimisation. Select assays which are fast and reliable.

### *Sample Extraction and Clarification*

The need for sample preparation prior to the first chromatographic step is dependent upon sample type. In some situations samples may be taken directly to the first capture step. For example cell culture supernatant can be applied directly to a suitable chromatographic matrix such as Sepharose™ Fast Flow and may require only a minor adjustment of the pH or ionic strength. However, it is most often essential to perform some form of sample extraction and clarification procedure. If sample extraction is required the chosen technique must be robust and suitable for all scales of purification likely to be used. It should be noted that a technique such as ammonium sulphate precipitation, commonly used in small scale, may be unsuitable for very large scale preparation.

Choice of buffers and additives must be carefully considered if a purification is to be scaled up. In these cases inexpensive buffers, such as acetate or citrate, are preferable to the more complex compositions used in the laboratory. It should also be noted that dialysis and other common methods used for adjustment of sample conditions are unsuitable for very large or very small samples.

— For repeated purification, use an extraction and clarification technique that is robust and able to handle sample variability. This ensures a reproducible product for the next purification step despite variability in starting material.

— Use additives only if essential for stabilisation of product or improved extraction. Select those which are easily removed. Additives may need to be removed in an extra purification step.

— Use pre-packed columns of Sephadex™ G-25 gel filtration media, for rapid sample clean-up at laboratory scale.

Sephadex G-25 gel filtration media are used at laboratory and production scale for sample preparation and clarification of proteins >5000. Sample volumes of up to 30%, or in some cases, 40% of the total column volume are loaded. In a single step, the sample is desalted, exchanged into a new buffer, and low molecular weight materials are removed. The high volume capacity, relative insensitivity to sample concentration, and speed of this step enable very large sample volumes to be processed rapidly and efficiently. Using a high sample volume load results in a separation with minimal sample dilution. Sephadex G-25 is also used for sample conditioning, e.g. rapid adjustment of pH, buffer exchange and desalting between purification steps.

*Combine sample clean-up and capture in a single step*

If large sample volumes will be handled or the method scaled-up in the future, consider using STREAMLINE™ expanded bed adsorption. This technique is particularly suited for large scale recombinant protein and monoclonal antibody purification. The crude sample containing particles can be applied to the expanded bed without filtration or centrifugation.

STREAMLINE adsorbents are specially designed for use in STREAMLINE columns. Together they enable the high flow rates needed for high productivity in industrial applications of fluidised beds. The technique requires no sample clean up and so combines sample preparation and capture in a single step. Crude sample is applied to an expanded bed of STREAMLINE media. Target proteins are captured whilst cell debris, cells, particulate matter, whole cells, and contaminants pass through. Flow is reversed and the target proteins are desorbed in the elution buffer.

**Capture**

— Use a high capacity, concentrating technique to reduce sample volume, to enable faster purification and to allow the use of smaller columns.

— Focus on robustness and simplicity in the first purification step. Do not try to solve all problems in one step when handling crude material.

Capture is the initial purification of the target molecule from crude or clarified source material. Main goals are rapid isolation, stabilisation and concentration. In the capture phase, the objective is to isolate, concentrate and stabilise the target product efficiently by optimising speed and capacity. The product is concentrated and transferred to an environment which will conserve activity. Capture is often a group separation using a step elution on ion exchange or affinity chromatography.

Ideally, removal of critical contaminants is also achieved. It is sometimes possible to achieve a high level of purification if a highly selective affinity media is used. Binding capacity for the protein in the presence of the impurities will be one of the most critical parameters to optimise and reduce the scale of work. For example, when ion exchange chromatography is used as a capture step, the goal is to adsorb the target protein quickly from the crude sample and isolate it from critical contaminants such as proteases and glycosidases.

Conditions are selected to avoid binding of contaminants so that the capacity for the target protein is maximised. High speed may be required to reduce sample application time, particularly if proteolysis or other destructive effects threaten the integrity of the target protein. The most common technique for a capture step is ion exchange chromatography (IEX) which has high binding capacity. IEX media are resistant to harsh cleaning conditions which may be needed after purification of crude samples. Typically proteins are eluted from an IEX column using a salt gradient.

However, during method development, a transfer to a step elution will give a simple, robust separation with a shorter run time and decreased buffer consumption. It is often possible to use high sample loadings since the focus is not on resolution (high sample loadings will decrease resolution).

High speed and capacity and low buffer consumption are particularly advantageous for large scale purification. For large scale capture, throughput will often be the focus during method development. It is important to consider all aspects: sample extraction and clarification, sample loading capacity, flow rate during equilibration, binding, washing, elution and cleaning, and the need for cleaning-in-place procedures. In principle, a capture step is designed to maximise capacity and/or speed at the expense of some resolution.

However, there is usually significant resolution and purification from molecules which have significant physicochemical differences compared to the target protein. Recovery will be of concern in any preparative situation, especially for production of a high value product, and it is important to assay for recovery during optimisation of the capture step.

### Intermediate Purification

In the intermediate purification phase the focus is to separate the target protein from most of the bulk impurities such as other proteins, nucleic acids, endotoxins and viruses. The ability to resolve similar components is of increased importance. The requirements for resolution will depend upon the status of the sample produced from the capture step and the purity requirements for the final product. Capacity will still be important to maintain productivity.

Speed is less critical in intermediate purification since the impurities causing proteolysis or other destructive effects should have been removed, and sample volume should have been reduced, in the capture step. The optimal balance between capacity and resolution must be defined for each specific application. This then decides how the separation conditions should be optimised during method development. The technique must give a high resolution separation. Elution by a continuous gradient will usually be required.

As in a capture step, selectivity during sample adsorption will be important, not only to achieve high binding capacity, but also to contribute to the purification by achieving a further separation during sample application. However, in contrast to a capture step, selectivity during sample desorption from the column is also important and is usually achieved by applying a more selective desorption principle, such as a continuous gradient or a multi-step elution procedure.

### Polishing

In the polishing phase the focus is almost entirely on high resolution to achieve final purity. Most contaminants and impurities have already been removed except for trace impurities such as leachables, endotoxins, nucleic acids or viruses, closely related substances such as microheterogeneous structural variants of the product, and reagents or aggregates. To achieve resolution it may be necessary to sacrifice sample load or even recovery (by peak cutting).

Recovery of the final product is also a high priority and a technique must be selected which ensures the highest possible recovery. Product losses at this stage are more costly than in earlier stages. Ideally the product should be recovered in buffer conditions ready for the next procedure. The technique chosen must discriminate between the target protein and any remaining contaminants

The high resolution required to achieve this discrimination is not always reached by using a high selectivity technique alone, but usually requires selection of a high efficiency medium with small, uniform bead sizes. Typically, separations by charge, hydrophobicity or affinity will have already been used so that high resolution gel filtration is ideal for polishing. The product is purified and transferred into the required buffer in one step and dimers or aggregates can often be removed. To remove contaminants of similar size, an alternative high resolution technique using elution with shallow gradients is usually required.

Final polishing step of recombinant epidermal growth factor, using reversed phase chromatography. Method developed on pre-packed RESOURCE™ RPC and scaled up on SOURCE 15RPC. Gel filtration is also the slowest of all chromatography techniques and the size of the column determines the volume of sample that can be applied. It is therefore most logical to use gel filtration after techniques which reduce sample volume so that smaller columns can be used.

When scaling up a purification it is important to verify that the high resolution achieved from the laboratory scale polishing step is maintained when applying preparative sample volumes to large scale columns.

### Selection and Combination of Purification Techniques

For any chromatographic separation each different technique will offer different performance with respect to recovery, resolution, speed and capacity. A technique can be optimised to focus on one of these parameters, for example resolution, or to achieve the best balance between two parameters, such as speed and capacity.

A separation optimised for one of these parameters will produce results quite different in appearance from those produced using the same technique, but focussed on an alternative parameter. Select a technique to meet the objectives for the purification step.

— *Capacity,* in the simple model shown, refers to the amount of target protein loaded during purification. In some cases the amount of sample which can be loaded may be limited by volume (as in gel filtration) or by large amounts of contaminants rather than the amount of the target protein.

— *Speed* is of the highest importance at the beginning of a purification where contaminants such as proteases must be removed as quickly as possible.

— *Recovery* becomes increasingly important as the purification proceeds because of the increased value of the purified product. Recovery is influenced by destructive processes in the sample and unfavourable conditions on the column.

— *Resolution* is achieved by the selectivity of the technique and the efficiency of the chromatographic matrix to produce narrow peaks. In general, resolution is most difficult to achieve in the final stages of purification when impurities and target protein are likely to have very similar properties.

Every technique offers a balance between resolution, speed, capacity and recovery and should be selected to meet the objectives for each purification step. In general, optimisation of any one of these four parameters can only be achieved at the expense of the others and a purification step will be a compromise.

The importance of each parameter will vary depending on whether a purification step is used for capture, intermediate purification or polishing. This will steer the optimisation of the critical parameters, as well as the selection of the most suitable media for the step. Proteins are purified using chromatographic purification techniques which separate according to differences in specific properties.

— Choose logical combinations of purification techniques based on the main benefits of the technique and the condition of the sample at the beginning or end of each step.

— Minimise sample handling between purification steps by combining techniques to avoid the need for sample conditioning.

If the sample has a low ionic strength it can be applied to an IEX column. After elution from IEX the sample will usually be in a high ionic strength buffer and can be applied to a HIC column (if necessary the pH can be adjusted and further salt can be added). In contrast, if sample is eluted from a HIC column, it is likely to be in high salt and will require dilution or a buffer exchange step in order to further decrease the ionic strength to a level suitable for IEX. Thus it is more straightforward to go from IEX to HIC than vice-versa.

Ammonium sulphate precipitation is a common sample clarification and concentration step at laboratory scale and in this situation HIC (which requires high salt to enhance binding to the media) is ideal as the capture step. The salt concentration and the total sample volume will be significantly reduced after elution from the HIC column. Dilution of the fractionated sample or rapid buffer exchange using a Sephadex G-25 desalting column will prepare it for the next IEX or AC step.

— GF is well suited for use after any of the concentrating techniques (IEX, HIC, AC) since the target protein will be eluted in a reduced volume and the components from the elution buffer will not affect the gel filtration separation.

— For any capture step, select the technique showing the strongest binding to the target protein while binding as few of the contaminants as possible, i.e. the technique with the highest selectivity and/or capacity for the protein of interest.

A sample is purified using a combination of techniques and alternative selectivities. For example, in an IEX-HIC-GF Three Phase Strategy the capture step selects according to differences in charge (IEX), the intermediate purification step according to differences in hydrophobicity (HIC) and the final polishing step according to differences in size (GF).

If nothing is known about the target protein use IEX-HIC-GF. This combination of techniques can be regarded as a standard protocol. Consider the use of both anion and cation exchange chromatography to give different selectivities within the same purification strategy. IEX is a technique which offers different selectivities using either anion or cation exchangers.

The pH of the separation can be modified to alter the charge characteristics of the sample components. It is therefore possible to use IEX more than once in a purification strategy, for capture, intermediate purification or polishing. IEX can be used effectively both for rapid separation in low resolution mode during capture, and in high resolution mode during polishing in the same purification scheme.

Consider Reversed phase chromatography (RPC) for a polishing step provided that the target protein can withstand the run conditions. RPC separates proteins and peptides on the basis of hydrophobicity. RPC is a high selectivity (high resolution) technique, requiring the use of organic solvents. The technique is widely used for purity check analyses when recovery of activity and tertiary structure are not essential. Since many proteins are denatured by organic solvents, the technique is not generally recommended for protein purification where recovery of activity and return to a correct tertiary structure may be compromised.

However, in the polishing phase, when the majority of protein impurities have been removed, RPC can be excellent, particularly for small target proteins which are often not denatured by organic solvents.

If a purification is not intended for scale up (i.e. only milligram quantities of product are needed), use high performance, pre-packed media such as Sepharose High Performance (IEX, HIC), SOURCE™ (IEX, HIC), MonoBeads™ (IEX), or Superdex (GF) for all steps.

### Sample Conditioning

Although additional sample handling between purification steps should be avoided, it may be necessary to adjust the buffer conditions of an eluted product (pH, ionic strength and/or buffering ions) to ensure compatibility with the following purification technique.

Sephadex G-25 is an ideal media for rapid desalting and pH adjustment by buffer exchange between purification steps. Sample volumes of up to 30%, or in some cases 40%, of the total column volume are loaded. In a single step, the sample is desalted, exchanged into a new buffer, and low molecular weight materials are removed. The high sample volume load results in a separation with minimal sample dilution. Sephadex G-25 is also used for rapid sample clean-up at laboratory scale.

### Principles and Conditions for Purification Techniques

With background information, assays, and sample preparation and extraction procedures in place the Three Phase Purification Strategy can be applied (Figure 2). This strategy is used as an aid to the development of purification processes for therapeutic proteins in the pharmaceutical industry and is equally efficient as an aid when developing purification schemes in the research laboratory.

*Assign a specific objective to each step within the purification process*: In the Three Phase Strategy a specific objective is assigned to each step. The purification problem associated with a particular step will depend greatly upon the properties of the starting material. Thus, the objective of a purification step will vary according to its position in the process, i.e. at the beginning for isolation of product from crude sample, in the middle for further purification of partially purified sample, or at the end for final clean up of an almost pure product. The Three Phase Strategy ensures faster method development, a shorter time to pure product and good economy.

In the capture phase the objectives are to isolate, concentrate and stabilise the target product. The product should be concentrated and transferred to an environment which will conserve potency/activity. At best, significant removal of other critical contaminants can also be achieved. During the intermediate purification phase the objectives are to remove most of the bulk impurities, such as other proteins and nucleic acids, endotoxins and viruses.

In the polishing phase most impurities have already been removed except for trace amounts or closely related substances. The objective is to achieve final purity. It should be noted that this Three Phase Strategy does not mean that all strategies must have three purification steps. For example, capture and intermediate purification may be achievable in a single step, as may intermediate purification and polishing. Similarly, purity demands may be so low that a rapid capture step is sufficient to achieve the desired result, or the purity of the starting material may be so high that only a polishing step is needed. For purification of therapeutic proteins a fourth or fifth purification step may be required to fulfil the highest purity and safety demands.

The optimum selection and combination of purification techniques for Capture, Intermediate Purification and Polishing is crucial for an efficient purification process.

### *Ion Exchange (IEX) Chromatography*

IEX separates proteins with differences in charge to give a very high resolution separation with high sample loading capacity. The separation is based on the reversible interaction between a charged protein and an oppositely charged chromatographic medium. Proteins bind as they are loaded onto a column. Conditions are then altered so that bound substances are eluted differentially. This elution is usually performed by increases in salt concentration or changes in pH. Changes are made stepwise or with a continuous gradient.

Most commonly, samples are eluted with salt (NaCl), using a gradient elution. Target proteins are concentrated during binding and collected in a purified, concentrated form. The net surface charge of proteins varies according to the surrounding pH. When above its isoelectric point (pI) a protein will bind to an anion exchanger, when below its pI a protein will behind to a cation exchanger. Typically IEX is used to bind the target molecule, but it can also be used to bind impurities if required. IEX can be repeated at different pH values to separate several proteins which have distinctly different charge properties. This can be used to advantage during a multi-step purification.

*Choice of Ion exchanger*

For most purification steps it is recommended to begin with a strong exchanger, allowing work over a broad pH range during method development. Use a strong anion exchanger (Q) to bind the target if the isoelectric point is below pH 7.0 or unknown.

— *Strong ion exchangers*: Q (anion exchange), S and SP (cation exchange) are fully charged over a broad pH range (pH 2 - 12).

— *Weak ion exchangers*: DEAE (anion exchange) and CM (cation exchange) are fully charged over a narrower pH range (pH 2 - 9 and pH 6 - 10, respectively), but give alternative selectivities for separations.

*Sample volume and capacity*

Ion exchange chromatography is a binding technique, independent of sample volume provided that the ionic strength of the sample is low and the target molecule is highly charged. The total amount of protein which is loaded and binds to the column should not exceed the total binding capacity of the column.

For optimal separations when performing gradient elution, use approximately one fifth of the total binding capacity of the column.

*Media selection*

Parameters such as scale of purification, resolution required, speed of separation, sample stability and media binding capacity, should be considered when selecting a chromatographic medium.

*Sample preparation*

Correct sample preparation ensures good resolution and extends the life of the column. To ensure efficient binding during sample application samples should be at the same pH and ionic strength as the starting buffer. Samples must be free from particulate matter, particularly when working with bead sizes of 34 μm or less.

*Column preparation*

— *Pre-packed columns*: To increase speed and efficiency in method development, use small pre-packed columns for media scouting and method optimisation. Using pre-packed columns at any scale will ensure reproducible results and high performance.

— *Column packing*: The following guidelines apply at all scales of operation:

  — Column dimensions = typically 5 - 15 cm bed height.

  — Quantity of gel = estimate amount of gel required to bind the sample, use five times this amount to pack a column.

*Buffer preparation*

Buffering ions should have the same charge as the selected medium, with a pKa within 0.6 pH units of the working pH. Buffer concentration should be sufficient to maintain buffering capacity and constant pH during sample application and while an increase in salt concentration is applied. When working with a sample of unknown charge characteristics, try these conditions first:

— Anion Exchange
  - — Gradient: 0-100% elution buffer B in 10 - 20 column volumes
  - — Start buffer A: 20 mM Tris-HCl, pH 8.0
  - — Elution buffer B: 20 mM Tris-HCl + 1 M NaCl, pH 8.0

— Cation Exchange
  - — Gradient: 0-100% elution buffer B in 10 - 20 column volumes
  - — Start buffer A: 20 mM $Na_2HPO_4.2H_2O$, pH 6.8
  - — Elution buffer B: 20 mM Na2HPO4.2H2O + 1 M NaCl, pH 6.8

***Hydrophobic Interaction Chromatography (HIC)***

HIC separates proteins with differences in hydrophobicity. The technique is ideal for the capture or intermediate steps in a purification. The separation is based on the reversible interaction between a protein and the hydrophobic surface of a chromatographic medium. This interaction is enhanced by high ionic strength buffer which makes HIC an ideal 'next step' after precipitation with ammonium sulphate or elution in high salt during IEX.

Samples in high ionic strength solution (e.g. 1.5 M ammonium sulphate) bind as they are loaded onto a column. Conditions are then altered so that the bound substances are eluted differentially. Elution is usually performed by decreases in salt concentration. Changes are made stepwise or with a continuous decreasing salt gradient. Most commonly, samples are eluted with a decreasing gradient of ammonium sulphate. Target proteins are concentrated during binding and collected in a purified, concentrated form. Other elution procedures include reducing eluent polarity (ethylene glycol gradient up to 50%), adding chaotropic species (urea, guanidine hydrochloride) or detergents, changing pH or temperature.

### *Hydrophobic ligand choice*

Very hydrophobic proteins bind tightly to very hydrophobic ligands and may require extreme elution conditions, e.g. chaotropic agents or detergents, for the target protein or contaminants. To avoid this problem it is recommended to screen several hydrophobic media, using HiTrap HIC Test Kit or RESOURCE HIC Test Kit. Begin with a medium of low hydrophobicity if the sample has very hydrophobic components. Select the medium which gives the best resolution and loading capacity at a reasonably low salt concentration.

Typically the strength of binding of a ligand to a protein increases in the order: ether, isopropyl, butyl, octyl, phenyl. However, the nature of the binding, both the selectivity and the binding strength, can vary and must be tested in individual cases.

### *Sample volume and capacity*

HIC is a binding technique and therefore rather independent of sample volume, provided that conditions are chosen to bind the target protein strongly. The total amount of protein which is loaded and binds to the column should not exceed the total binding capacity of the column. For optimal separations when performing gradient elution, use approximately one fifth of the total binding capacity of the column

In HIC the characteristics of the chromatographic matrix as well as the hydrophobic ligand affect the selectivity of the medium. This should be considered, together with parameters such as sample solubility, required resolution, scale of purification and availability of the medium at the scale intended.

*Sample preparation*

Correct sample preparation ensures good resolution and extends the life of the column. To ensure efficient binding during sample application samples should be at the same pH as the starting buffer and in high ionic strength solution (e.g. 1.5 M ammonium sulphate or 4 M NaCl). Samples must be free from particulate matter, particularly when working with bead sizes of 34 μm or less.

*Column Preparation*

*Pre-packed columns:* To increase speed and efficiency in method development use small pre-packed columns for media scouting and method optimisation. HiTrap HIC Test Kit and RESOURCE HIC Test Kit are ideal for this method. Using pre-packed columns at any scale will ensure reproducible results and high performance.

*Column packing:* The following guidelines apply at all scales of operation:

— Column dimensions = typically 5 - 15 cm bed height.

— Quantity of gel = estimate amount of gel required to bind the sample, use five times this amount to pack a column.

*Buffer preparation*

Buffering ion selection is not critical for hydrophobic interaction. Select a pH compatible with protein stability and activity. Buffer concentration must be sufficient to maintain pH during sample application and changes in salt concentration. When working with a sample of unknown hydrophobic characteristics, try these conditions first:

— Gradient: 0-100% elution buffer B in 10 - 20 column volumes

— Start buffer A: 50 mM sodium phosphate pH 7.0 + 1 - 1.5 M ammonium sulphate

— Elution buffer B: 50 mM sodium phosphate pH 7.0

*Method development*

1) The hydrophobic behaviour of a protein is difficult to predict and binding conditions mst be studied carefully. Use a HiTrap HIC Text Kit or a RESOURCE HIC Test Kit to select the medium which gives optimum binding and elution over the required range of salt concentration. For proteins with unknown hydrophobic properties begin with 0-100%B (0%B=1 M ammonium sulphate).

2) Select the gradient which gives acceptable resolution.

3) Select the highest flow rate which maintains resolution and minimises separation time. Check recommended flow rates for the specific medium.

4) For large scale purification and to reduce separation times and buffer consumption, transfer to a step elution after method optimisation. It is often possible to increase sample loading when using step elution, an additional benefit for large scale purification.

5) If samples adsorb strongly to a gel then conditions which cause conformational changes, such as pH, temperature, chaotropic ions or organic solvents can be altered. Conformational changes caused by these agents are specific to each protein. Use screening

procedures to investigate the effects of these agents. Alternatively, change to a less hydrophobic medium.

### *Affinity Chromatography (AC)*

AC separates proteins on the basis of a reversible interaction between a protein (or group of proteins) and a specific ligand attached to a chromatographic matrix. The technique is ideal for a capture or intermediate step and can be used whenever a suitable ligand is available for the protein(s) of interest. AC offers high selectivity, hence high resolution, and usually high capacity for the protein(s) of interest. The target protein(s) is specifically and reversibly bound by a complementary binding substance (ligand).

The sample is applied under conditions that favour specific binding to the ligand. Unbound material is washed away, and the bound target protein is recovered by changing conditions to those favouring desorption. Desorption is performed specifically, using a competitive ligand, or non specifically, by changing the pH, ionic strength or polarity. Samples are concentrated during binding and protein is collected in purified, concentrated form. Affinity chromatography is also used to remove specific contaminants.

*Sample volume and capacity*: AC is a binding technique, independent of sample volume provided that conditions are chosen to bind the target protein strongly.

#### *Media selection*

Parameters such as scale of purification and commercial availability of affinity matrices should be considered when selecting affinity media. To save time and ensure reproducibility, use prepacked columns for method development or small scale purification.

*Sample preparation*

Correct sample preparation ensures efficient binding and extends the life of a column. Removal of contaminants which may bind non-specifically to the column, such as lipids, is crucial. Stringent washing procedures may damage the ligand of an affinity medium, destroying the binding capacity of the column. Samples must be free from particulate matter.

*Column preparation*

— *Pre-packed columns:* Pre-packed columns ensure reproducible results and highest performance.

— *Column packing:* The following guidelines apply at all scales of operation:

  — Column dimensions = short and wide.

  — Quantity of gel = calculate according to known binding capacity of medium, use 2-5 times excess capacity.

*Buffer preparation*

Binding, elution and regeneration buffers are specific to each affinity medium. Follow instructions supplied with the medium or column.

1) Select the correct specificity for the target protein. Follow the manufacturer's instructions for binding or elution conditions and check recommended flow rates for the specific medium.

2) Select optimum flow rate to achieve efficient binding

3) Select optimum flow rate for elution to maximise recovery.

4) Select maximum flow rate for column regeneration to minimise run times.

Cleaning, sanitisation and *sterilisation*: Procedures vary according to type of sample and medium. Guidelines are supplied with the medium or pre-packed column.

*Storage of media and columns*: Follow the manufacturer's instructions.

### *Gel Filtration (GF)*

GF separates proteins with differences in molecular size. The technique is ideal for the final polishing steps in a purification when sample volumes have been reduced (sample volume significantly influences speed and resolution in gel filtration). Samples are eluted isocratically. Buffer conditions are varied to suit the sample type or the requirements for further purification, analysis or storage step, since buffer composition does not directly affect resolution. Proteins are collected in purified form in the chosen buffer.

#### *Sample volume and capacity*

To achieve the highest resolution the sample volume must not exceed 5% of the total column volume. Gel filtration is independent of sample concentration, although above 50 mg/ml protein viscosity effects may cause 'fingering'. Extremely viscous samples should be diluted.

#### *Media selection*

Parameters such as molecular weight of target proteins and contaminants, resolution required, scale of purification should be considered when selecting gel filtration media.

#### *Sample Preparation.*

Correct sample preparation ensures good resolution and extends the life of the column. Sample buffer composition does not directly affect resolution. During separation the sample buffer is exchanged with buffer in the column. Viscous samples, which could cause an increase in back pressure and affect column packing, should be diluted. Samples must be free from particulate matter, particularly when working with bead sizes of 34 μm or less.

*Column preparation*

— *Pre-packed columns*: Pre-packed columns ensure reproducible results and highest performance.

— *Column packing*: In gel filtration good column packing is essential. The resolution between two separated zones increases as the square root of column length. The following guidelines apply:

  — Column dimensions: = minimum 50 cm bed height (Sephacryl) minimum 30 cm bed height (Superdex, Superose)

  — Bed volume = depending on sample volume per run (up to 5% of bed volume)

*Buffer preparation*

Selection of buffering ion does not directly affect resolution. Select a buffer in which the purified product should be collected and which is compatible with protein stability and activity. Buffer concentration must be sufficient to maintain buffering capacity and constant pH. Ionic strength can be up to 150 mM NaCl in the buffer, to avoid non-specific ionic interactions with the matrix.

*Method development*

1. Select the medium which gives the best separation of target proteins from contaminants.

2. Select the highest flow rate which maintains resolution and minimises separation time. Check recommended

flow rates for the specific medium. Lower flow rates improve resolution of high molecular weight components, faster flow rates may improve resolution of low molecular weight components

3. Determine the maximum sample volume which can be loaded without reducing resolution (sample volume should be 0.5-5% of total column volume).

4. To further improve resolution increase column length by connecting two columns in series

*Cleaning, sanitisation and sterilisation*: Procedures vary according to type of sample and medium. Guidelines are supplied with the medium or pre-packed column.

*Storage of media and columns*: Recommended conditions for storage are supplied with the medium or pre-packed column.

#### Reversed Phase Chromatography (RPC)

RPC separates proteins and peptides with differing hydrophobicity based on their reversible interaction with the hydrophobic surface of a chromatographic medium. Samples bind as they are loaded onto a column. Conditions are then altered so that the bound substances are eluted differentially. Due to the nature of the reversed phase matrices, the binding is usually very strong and requires the use of organic solvents and other additives (ion pairing agents) for elution.

Elution is usually performed by increases in organic solvent concentration, most commonly acetonitrile. Samples, which are concentrated during the binding and separation process, are collected in a purified, concentrated form.

RPC is often used in the final polishing of oligonucleotides and peptides and is ideal for analytical separations, such as peptide mapping. RPC is not recommended for protein purification if recovery of activity and return to a correct tertiary structure are required, since many proteins are denatured in the presence of organic solvents.

*Choice of hydrophobic ligand*

Select hydrocarbon ligands according to the degree of hydrophobicity required. Highly hydrophobic molecules bind tightly to highly hydrophobic ligands, e.g. C18.

Screen several RPC media. Begin with a medium of low hydrophobicity, if the sample has very hydrophobic components (more likely with larger biomolecules, such as proteins). Select the medium which gives the best resolution and loading capacity. A polymer based medium such as SOURCE RPC can offer significant advantages over silica based media as it can be used across the pH range 1-14 providing not only an alternative selectivity to silica but also a wider working pH range for method optimisation.

*Sample volume and capacity*

RPC is a binding technique, often independent of sample volume. Total capacity is strongly dependent upon experimental conditions and the properties of the gel and sample. For optimal conditions during gradient elution, screen for a sample loading which does not reduce resolution.

*Media selection*: In RPC the chromatographic medium as well as the hydrophobic ligand affect selectivity. Screening of different RPC media is recommended.

*Sample preparation*: Samples should be free from particulate matter and, when possible, dissolved in the start buffer.

*Column preparation*: Reversed phase columns should be 'conditioned' for first time use, after long term storage or when changing buffer systems.

*Buffer preparation*: Try these conditions first when sample characteristics are unknown:

— Gradient: 2-80% elution buffer B in 20 column volumes

— Start buffer A: 0.065% TFA (trifluoroacetic acid) in water

— Elution buffer B: 0.05% TFA in acetonitrile

*Method development*

1. Select medium from screening results.
2. Select gradient to give acceptable resolution. For unknown samples begin 0-100%B.
3. Select the highest flow rate which maintains resolution and minimises separation time.
4. For large scale purification transfer to a step elution.
5. Samples which adsorb strongly to a gel are more easily eluted from a less hydrophobic medium.

*Cleaning, sanitisation and sterilisation*: Procedures vary according to type of sample and medium. Guidelines are supplied with the medium or pre-packed column.

*Storage of media and columns*: Recommended conditions for storage are supplied with the medium or pre-packed column.

#### Expanded Bed Adsorption (EBA)

EBA is a single pass operation in which target proteins are purified from crude sample, without the need for separate clarification, concentration and initial purification to remove particulate matter. Crude sample is applied to an expanded bed of STREAMLINE adsorbent particles within a specifically designed STREAMLINE column. Target proteins are captured on the adsorbent. Cell debris, particulate matter, whole cells, and contaminants pass through and target proteins are then eluted.

### *Selection of STREAMLINE adsorbent*

Selection of adsorbent is based on the same principles that are used for chromatography. Select the medium with the strongest binding to the target protein and which binds as few of the contaminants as possible, i.e. the medium with the highest selectivity and/or capacity for the protein of interest.

### *Sample volume and capacity*

All STREAMLINE media are binding techniques, independent of sample volume. The total amount of protein which is loaded should not exceed the total binding capacity of the column.

### *Sample preparation*

STREAMLINE is able to handle crude, particulate feedstock, reducing the need for significant sample preparation steps. Adjustment of pH or ionic strength may be required according to the separation principle being used (IEX, AC, HIC)

### *Column preparation*

For preliminary method scouting STREAMLINE media is used in packed bed mode in an XK 16 or XK 26 chromatography column. When used in expanded bed mode the media must be packed in specially designed STREAMLINE columns, following the manufacturer's instructions.

*Buffer Preparation*: Buffer preparation will depend upon the chosen separation principle.

*Method development*

1) Select suitable ligand to bind the target protein.

2) Scout for optimal binding and elution conditions using clarified material in a packed column (0.02 - 0.15 litres bed volume of media). Gradient elutions may be used during scouting, but the goal is to develop a step elution.

3) Optimise binding, elution, wash and cleaning-in-place procedures using unclarified sample in expanded mode at small scale (0.02 - 0.15 litres bed volume of media)

4) Begin scale up process at pilot scale (0.2 - 0.9 litres bed volume of media)

5) Full scale production (up to several hundred litres bed volume of media)

*Cleaning, sanitisation and sterilisation*: Guidelines are supplied with each STREAMLINE adsorbent.

# 5

# Protein Crystal Technology

Biocatalysts are commonly used as impure water-soluble protein preparations that contain only a small fraction of the active catalyst (usually 0.5-10%). In catalysis it is even possible that more catalyst is added than reactant. The properties of enzymes have been improved by protein engineering or by immobilisation, for example. Protein engineering can be used in order to enhance both stability and activity. However, protein engineering is laborious and only one property of one protein can be altered at one time. Furthermore, expression systems and an economically viable production process have to be developed for the new protein.

Enzyme immobilisation is used to facilitate the downstream processing and to improve the stability of enzymes. Immobilised enzymes can be easily removed from the reaction mixture and, thus, can be recycled and reused. However, the stability enhancement is usually negligible. Moreover, the use of an inert solid support leads to a low overall activity of an immobilised protein matrix. Chemical cross-linking of enzyme crystals was first developed in the 1960's to stabilise enzyme crystals in order to facilitate the manipulation and the collection of crystallographic data by X-ray crystallography. However, it was soon recognised that cross-linked crystalline enzymes were catalytically active.

Cross-linked enzyme crystals, or more generally cross-linked protein crystals (CLPCs) are created by crystallisation of protein followed by a chemical cross-linking that "locks" the crystalline state outside the crystallisation conditions. CLPCs form an active, insoluble, mechanically robust, and microporous protein matrix. CLPCs have been used as catalysts in various enzymatic production processes, as separation matrices, and in some other applications. However, the development of CLPC technology is still under way and the number of proteins commercially available as cross-linked crystals is very limited. Protein crystallisation and CLPC technology are described below in greater detail.

#### Protein Crystallisation

A solution system remains at equilibrium until the point is reached where there is insufficient solvent to maintain full hydration of the solute molecules. If more solute is added to the solution, the system is no longer at equilibrium under this so-called supersaturated state. The system will be thermodynamically driven toward a new equilibrium corresponding free energy minimum by forming a solid phase. The solid phase can appear as either amorphous precipitate or crystals. The principle of crystallisation is similar for small molecules (e.g., salts and small organic compounds) and macromolecules (e.g., proteins, DNA, and RNA). Three stages are common to all crystallisations: nucleation, crystal growth, and cessation of crystal growth.

In nucleation molecules or non-crystalline aggregates (dimers, trimers, etc.) produce a stable aggregate with a repeating structure. The nucleus must first exceed a specific size, called the critical size, before it is capable of further growth. However, the formation of crystal nuclei from supersaturated solution does not necessarily result in the subsequent formation of macroscopic crystals. Cessation of crystal growth can occur for many reasons.

The most obvious reason is that the equilibrium between the crystalline and soluble form is achieved. The crystallisation of proteins is usually more difficult than small molecule crystallisation. This is probably a result of the difficulty to obtain a high quality protein sample for crystallisation. Because precise protein-protein contacts are needed for the crystallisation of proteins, any factor that introduce heterogeneity into the protein sample may have interfering effects on the crystallisation.

There are numerous factors that can introduce heterogeneity into a protein sample including differential glycosylation on the protein molecules, different conformations of the protein can be present, contamination in the form of other proteins, protein denaturation, and protein degradation. The basic strategy to crystallise proteins is to bring the system into a state of limited degree of supersaturation. This can be done, for example, by removing the solvent, adding a precipitating agent, or by altering some physical properties such as temperature. However, there are dozens of other parameters that are involved in protein crystallisation. Protein crystallisation is presented in detail in numerous handbooks.

#### Uses of Protein Crystallisation

Protein crystallisation is mainly used for structure determination by X-ray crystallography where large, high quality crystals are needed. Obtaining high quality diffractive crystals is the bottleneck in protein structure determination. For structure determination purposes protein crystals are most often crystallised by small-scale microdiffusion methods. Very often crystallisation conditions are found after setting up hundreds or thousands of individual crystallisation conditions employing the "trial and error" approach, as crystallisation is often considered as a necessary evil in the structure determination work.

The structural analysis is based on the principle that in a perfect crystal all the molecules have the same conformation and orientation. When X-rays are focused through protein (or any other pure substance) crystal, individual atoms diffract the rays. The number of electrons in each atom determines the intensity of the scattering of X-rays and, thus, the X-rays striking particular atoms will all be diffracted in the same way. A diffraction pattern is generated by the interference of individual X-rays. A series of diffraction patterns are taken from several angles and these patterns represent the way atoms are arranged in the molecule and can be used to determine the structure of the protein. There are currently more than 18000 protein structures in the Protein Data Bank (PDB), which have been determined by X-ray diffraction analysis.

Crystallisation is an efficient protein purification method. However, purification by crystallisation is relatively rare even though the quality of the crystals is not crucial. Chromatographic methods have replaced protein crystallisation in protein purification although the latter was commonly used earlier. However, protein purification by crystallisation has many advantages: high yield, high purity in one step, unlimited scaleup possibilities, and the product is a highly concentrated protein crystal slurry ready for further formulation. Some industrial proteins, like xylose isomerase, cellulase, and protease, have been purified by crystallisation in large-scale.

A majority of small molecular weight drugs are produced in crystalline form because of the high storage stability, purity, and reproducibility of the drug properties. There are hundreds of macromolecular therapeutic agents used in clinical trials or approved as drugs. However, only insulin is produced and administered in a crystalline form. According to Margolin and Navia the crystallisation of macromolecular pharmaceuticals can offer significant advantages, such as:

a) protein purification by crystallisation as presented above;

b) high stability of the protein product compared with soluble or amorphous forms;

c) crystals are the most concentrated form of proteins, which is a benefit for storage, formulation, and for drugs that are needed in high doses (e.g., antibiotics);

d) the rate of crystal dissolution depends on the morphology, size, and additives; thus, crystalline proteins may be used as a carrier-free dosage form.

### *Cross-Linked Protein Crystals (CLPCs)*

In many applications crystallised proteins are not suitable for use as such as a result of their fragility and solubility. In order to produce a crystalline protein matrix that is insoluble also in other conditions than those used in crystallisation, the crystals have to be chemically cross-linked. In general, chemical cross-linking of protein crystals creates an active and microporous protein matrix that can be used in catalytic and separation applications.

#### *Glutaraldehyde*

Protein crystals dissolve in other conditions than those used in the crystallisation. Chemical cross-linking of protein crystals "locks" the crystalline state and, thus, the cross-linked protein crystals are insoluble also outside the conditions that initially led to crystallisation. The most commonly used cross-linking reagent is a bifunctional aldehyde, glutaraldehyde.

Most often, the cross-linking reaction is carried out near neutral pH by adding 0.1-2% (w/v) glutaraldehyde to the crystallisation mother liquor containing the native protein crystals. This mixture is then incubated in cool conditions (0-25 °C) for 0.5-2 h. The resulting chemical cross-linking takes place throughout the crystalline lattice and is intermolecular but intracrystalline.

It is well established that glutaraldehyde reacts with the free ε-amino groups of lysine residues in proteins. In numerous studies the amino acid composition of cross-linked protein crystals has been analysed and it was clearly shown that only lysine residues of proteins reacted in the cross-linking reaction by glutaraldehyde. As an exception, an amino acid analysis of cross-linked chloroperoxidase crystals suggested that also serines and arginines were modified in the glutaraldehyde cross-linking reaction. Crystallographic studies show that glutaraldehyde crosslinking reaction had only minor effects on the structure of lysozyme, carboxypeptidase A, subtilisin, an N-cadherin fragment, and a gp120 ternary complex.

The X-ray crystallographic data from triclinic lysozyme crystals indicate that only lysine residues were modified in glutaraldehyde cross-linking. In principle, the carbonyl function in aldehydes reacts with free amino groups and the reaction product is a Schiff base. A general characteristic of Shiff bases is that they are readily hydrolyzed in acidic conditions. However, numerous examples show that the cross-linking reaction between glutaraldehyde and proteins is irreversible, shown by incubating cross-linked protein crystals in acidic solution, for example. This rules out a simple Shiff base formation.

Glutaraldehyde is known to form polymers in solution. It is speculated that the ability of glutaraldehyde to stabilise protein crystals stems from the fact that in aqueous solution glutaraldehyde forms a mixture of oligomers of different lengths and structures, and therefore the most appropriate cross-linking species reacts with crystalline protein. At acidic pH, the oligomers are cyclic hemiacetals (Figure 1a) that form Shiff bases in analogy with the mechanism of Cordes and Jencks.

At neutral or slightly alkaline pH, where protein cross-linking is carried out, α, β-unsaturated glutaraldehyde polymers are formed (Figure 1b). Basing his conclusions on earlier studies, Wong proposed a reaction mechanism where the unsaturated glutaraldehyde polymer cross-links the amino groups (lysine residues) of proteins as shown in Figure 1b.

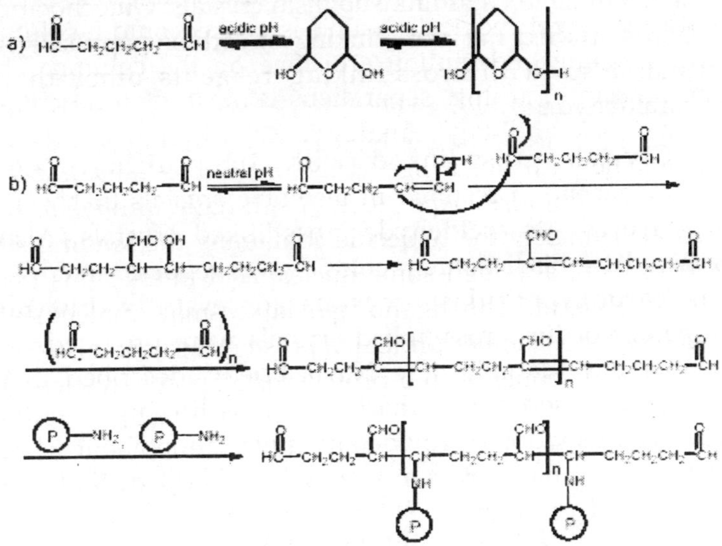

*Figure 1. Polymerisation of glutaraldehyde at a) acidic pH and b) at neutral pH and its cross-linking reaction with proteins.*

Rydon and co-workers have proposed a different reaction mechanism, where glutaraldehyde cross-linking of proteins leads mainly to formation of stable quaternary pyridinium compounds. However, when an amino group containing compound was added to the cross-linking reaction mixture, the resulting cross-linked XI crystals were insoluble. Visuri used ammonium salts, amines, and amino acids as the source of the additional amino groups. The reason for the improved cross-linking in the presence of exogenous amino groups remained unknown.

### *Other Cross-linking Reagents*

A great number of chemical cross-linking reagents capable of reacting with the side chains of amino acids of proteins have been identified. However, in addition to glutaraldehyde, only a few of them have been used in the preparation of cross-linked protein crystals. Dialdehydes analogous to glutaraldehyde have been used to a minor extent. Margolin et al. used succinaldehyde, glyoxal, and octanaldehyde to cross-link subtilisin crystals. Quiocho and Richards studied the cross-linking of carboxypeptidase A crystals also with cross-linking reagents other than glutaraldehyde.

Crystals cross-linked with 1,3- difluoro-4,6-dinitrobenzene in resulted in insoluble crystals that were as active as glutaraldehyde cross-linked crystals. Also biphenyl-4,4'-bisdiazonium chloride was reported to cross-link carboxypeptidase crystals successfully, but no properties of the cross-linked crystals were provided. In their review Margolin and Navia briefly described that they have used more than ten cross-linking agents. However, they don't provide any information about what proteins they have cross-linked and the properties of the cross-linked products.

Margolin and Navia described also a reversible crosslinking method. They modified Candida rugosa lipase by Traut's reagent in order to introduce additional SH groups to the protein. Then the SH groups were further cross-linked with DPDPB.

When the water-insoluble cross-linked product was incubated with exogenous cysteine, the crystals dissolved completely and released 85% of the initial lipase activity. Margolin et al. described the cross-linking of glycoprotein crystals based on chemical carbohydrate-to-carbohydrate cross-linking. Carbohydrates in glucose oxidase crystals were first oxidised by sodium periodate and then cross-linked by 1,8-diaminooctane.

Thereafter $NaCHBH_3$ was used to reduce the imino groups in the product. The resulting cross-linked product was slightly soluble in water and exhibited specific activity similar to the soluble counterpart. These crystals were used to produce gluconic acid. Margolin et al. also prepared carbohydrate cross-linked crystals of C. rugosa lipase and several vaccine antigens.

### *Porous Structure of CLPCs*

In a crystalline form proteins are in a highly ordered three-dimensional array where the protein molecules are bound to each other with specific intermolecular interactions. Protein crystals contain not only protein molecules but also uniform solvent-filled pores that constitute 30-78% of the crystal volume depending on the protein and the crystallisation conditions. The solvent flows almost freely in the pores of crystalline proteins.

Morozov et al. measured that the self-diffusion coefficient of intracrystalline water in the tetragonal lysozyme crystals was reduced by about 30-40% compared with that of bulk water. Table 1 shows that the average diameter of the pores in cross-linked crystals varies between 25-80Å, the porosity is 0.5-0.8, and the pore surface area is 600-1800 $m^2$ $g^{-1}$ depending on the protein in question.

The pore surface area of cross-linked crystals is comparable with that of inorganic porous materials, such as zeolites, silica gels, and activated carbon. The pore sizes in Table 1 have been calculated by using a cylindrical pore model and symmetrical pore distribution.

*Table 1. Characteristics of CLPCs.*

| Protein | Solvent content | Porosity | Pore surface area ($m^2$ $g^{-1}$) | Pore diameter (Å) |
|---|---|---|---|---|
| Thermolysin | 50% | 0.51 | 800 | 46 |
| Lipase (Pseudomonas cepacia) | 74% | 0.80 | 1800 | 82 |
| Lipase (C. rugosa) | 49% | 0.50 | 900 | 38 |
| Human serum albumin | – | 0.65 | 1500 | 54 |
| Xylose isomerase | 41–53% | 0.47 | 600 | 29 |

*Figure 2. Crystal lattice structure of Trichoderma reesei xylanase II crystal.*

However, the pores within protein crystals are complicated in structure and cannot, thus, be described by a simple cylindrical model.

Morozova et al. presented that the pore size in cross-linked tetragonal lysozyme crystals varied between 6-40 Å measured from crystallographic data. Figure 2 illustrates the pore structure in a native xylanase (EC 3.2.1.8) crystal lattice. The pore sizes measured from crystallographic data are systematically smaller than presented in table 1.

### Activities of CLPCs

Mass-transfer limitations have been recognised in numerous publications to decrease the activity of CLPCs compared with the soluble counterparts. There are three factors affecting the mass-transfer limitations: the size of the crystalline particle, the size of the substrate, and the pore size of the crystal. As shown above, the sizes of the open channels in protein crystals are 10-80 Å in diameter.

As a consequence, large molecules having a molecular weight higher than 1000-10000 g $mol^{-1}$, depending on the protein crystal in question, cannot penetrate the crystals at all. Smaller molecules can diffuse through the pores. However, the diffusion rate can be considerably decreased compared with the diffusion rate in water. The effective diffusion coefficients of surfactants (MW 246-316 g $mol^{-1}$) inside tetragonal lysozyme crystals were 3-4 orders of magnitude lower than those of the corresponding free molecules in water.

Morozova et al. calculated that the mobility of cations was 4-50 fold lower and that of anions 100-300 fold lower in the tetragonal lysozyme crystals compared with the mobilities in solution. Even the self-diffusion coefficient of intracrystalline water in the tetragonal lysozyme crystal is reduced by about 30-40% compared to that of bulk water. The crystal size plays a crucial role in the activity of the crystals. Most of the catalytic reactions are taking place inside the crystals, as even with very small crystals (e.g., 1×1×2 μm³) more than 99% of the protein molecules are located in the interior of the crystal.

As a consequence, the substrate and the product have to diffuse from the bulk to the catalytic site and from the catalytic site to bulk solution, respectively. Table 2 shows the relative activity of cross-linked crystalline enzymes in respect to the crystal size. The effect of the crystal size on the activity depends on the kinetics of the reaction, the size of the pores, and the size of the substrate and product. The crystal size has the most dramatic effect on reactions involving large molecules that can diffuse only partially or cannot diffuse at all to the pores of crystal.

*Table 2. The effect of particle size on the relative activity of cross-linked protein crystals.*

| Protein | Particle size in the smallest dimension (μm) | Relative activity (per unit weight of enzyme) |
|---|---|---|
| Carboxypeptidase A | 58 | 1 |
| | 28 | 1.6 |
| | 6 | 5.6 |
| Alcalase | 40 | 1 |
| (EC 3.4.21.-) | 1.2 | 3.2 |
| Penicillin acylase | 142 | 1 |
| (EC 3.5.1.11) | 57 | 3.4 |

Limited conformational flexibility may cause dramatically reduced protein activities. In the crystal lattice proteins have multiple specific intermolecular interactions. These interactions may stabilise proteins but can also reduce the flexibility of protein molecules. Another source of limited flexibility is the chemical cross-linking. In recent reports Ayala et al. and Costes et al. showed that the specific activity of crosslinked crystals of chloroperoxidase and hydroxynitrile lyase (EC 4.1.2.37), respectively, decreased with increasing the glutaraldehyde concentration used in cross-linking. However, maximal stability and total insolubility was only achieved in conditions using high concentration of glutaraldehyde in both cases. Similar results were obtained with subtilisin.

However, the reduced protein flexibility caused by crystallisation and cross-linking can also be exploited. The properties of cross-linked C. rugosa lipase crystal preparations were affected by the conformation of lipase during crystallisation and cross-linking. In lipase molecule there is a certain "lid" that is "open" enabling substrate binding to the active site when the active site is unoccupied. However, when the "lid" is "closed" the substrate is enclosed to the active site of lipase. If lipase was crystallised and cross-linked "lid open", the activity was three times higher than if crystallised "lid closed".

The crystallisation and cross-linking of lipase in "lid open" form allowed a rapid binding of the substrate to the active site and a rapid removal of the product from the active site thus increasing the specific activity. In addition to the reduced flexibility the cross-linking can also have other effects on the activity of protein crystals. It is possible that glutaraldehyde cross-linking creates steric hindrances in the crystal lattice, which may occlude some active sites. Chemical cross-linking may also modify catalytically important residues at or near the active site and, thus, affect the catalytic properties of the protein.

The chemical cross-linking of protein crystals significantly enhances the stability against heat denaturation, proteolytic enzymes, and organic solvents compared with corresponding soluble proteins. The enhanced stability can be a result of crystallinity, chemical cross-linking of the proteins, or a combination of both. In the crystalline form proteins maintain better their native conformation in conditions that would lead to an unfolding of soluble protein. This is a result of the highly ordered three-dimensional array of protein molecules in crystal where protein molecules are bound to each other with specific intermolecular interactions.

The crystalline environment enhances the stability of proteins against heat and other denaturants by preventing unfolding, aggregation, or dissociation of proteins. The chemical cross-linking is another source of stabilisation. It has been shown that long glutaraldehyde cross-linking times yielding a high degree of cross-linking is needed to achieve maximal stability of cross-linked protein preparates. It seems that cross-linked crystals are universally resistant to digestion by proteolytic enzyme. Possibly, proteases can, at least partly, digest the proteins locating on the protein crystal surface. However, these digested surface proteins remain attached to the surface of the crystal preventing the proteases to enter the crystals. Because only a small fraction of the protein molecules are located on the crystal surface, the activity of crosslinked crystal remains at the initial level.

Native protein crystals are known to be fragile and they are easily cracked or chipped even by gentle mechanical forces. In contrast, cross-linked protein crystals are considerably more stable towards mechanical stresses than native crystals. The source of this mechanical stabilisation is clearly the chemical cross-linking. Quite surprisingly, only Lye and co-workers have actually measured the mechanical stability of cross-linked protein crystals using C. rugosa lipase and yeast alcohol dehydrogenase as model crystals. Flat plate (thickness <2 μm) crystals of lipase cross-linked in 0.5% glutaraldehyde and yeast alcohol dehydrogenase cross-linked in 0.05% glutaraldehyde remained intact in conditions equivalent to those expected in the highshear regions of stirred-tank reactors ($\varepsilon_{max}$ = 0.34 ×$10^5$ W $kg^{-1}$).

The native crystals of yeast alcohol dehydrogenase were significantly fragmented in the same conditions. However, at high shear stress ($\varepsilon_{max}$ = 6.8 ×$10^5$ W $kg^{-1}$) both lipase and yeast alcohol dehydrogenase crystals were broken, and the median crystal size was decreased from an initial value of 13 µm to 7 µm and 12 µm to 7 µm for lipase and yeast alcohol dehydrogenase, respectively. The lipase crystals that were cross-linked in 5% glutaraldehyde, and small cross-linked rod-shaped yeast alcohol dehydrogenase crystals having an average size of 5 µm, were totally resistant even to the high shear stress ($\varepsilon_{max}$ = 6.8 ×$10^5$ W $kg^{-1}$). The visual inspection of the broken crystals showed significant rounding of the crystal edges. This suggests that the mechanism of breakage is primarily a result of attrition by crystal-crystal collisions.

The mechanical integrity of cross-linked yeast alcohol dehydrogenase crystals was studied under compressive and abrasive forces found in dead-end filtration. The cake resistance was constant over five cycles of mixing and filtration at a pressure of 100 kPa indicating no breakage of the crystals. Rod-shaped crystals had a lower cake resistance than the hexagonal crystals. This was explained by the size and shape of the crystals. Rod-shaped crystals are cylindrical and allow more voidage within the cake. The hexagonal crystals are flat plates that can stack on each other forming a very dense cake increasing the cake resistance.

The reason for the choice to use an enzyme in the form of cross-linked crystals is most often based on the stability of the cross-linked crystals in both aqueous and organic media. The stability is achieved without a considerable loss of specific activity. Crosslinked crystals are insoluble throughout the production process. They are mechanically robust and can be easily removed from the reaction mixture by filtration or centrifugation and, thus, they can be recycled, which increases the productivity of the catalyst. The purity of the catalyst leads to a high enantioselectivity of the reaction. Cross-linked crystals are, in addition, an extremely stable storage formula compared to soluble enzymes, for example. Table 3 summarises the catalytic applications where cross-linked enzyme crystals have been used.

*Table 3. Catalytic applications where cross-linked enzyme crystals have been used.*

| Enzyme | Reaction |
|---|---|
| Subtilisin | Synthesis of peptides and peptidomimetics |
| Subtilisin | Regioselective hydrolysis |
| Subtilisin | Transesterification |
| Subtilisin | Catalysis in supercrirical fluids |
| Thermolysin | Synthesis of peptides |
| Lipase | Enantioselective ester hydrolysis |
| Lipase | Enantioselective transesterification |
| Lipase | Enantioselective acylation |
| Hydroxynitrile lyase | Asymmetric syntheses |
| Fructose diphosphate aldolase | Synthesis of tagetitoxin |
| Xylose isomerase | Carbohydrate conversions |
| Xylanase | Hydrolysis of xylo-oligomers |
| Alcohol dehydrogenase | Reduction of ketones with cofactor regeneration |
| Lactate dehydrogenase | Production of lactate |
| Organophosphorous hydrolase (EC 3.1.8.1) | Hydrolysis of pesticide |

***Proteases***

The protease subtilisin is used in synthetic chemistry in both aqueous and organic solvents for regio- and enantiospecific reactions, amide bond syntheses, and hydrolyses. Two major problems have limited the use of subtilisin: low stability in water and watermiscible solvents, and low activity in neat organic solvents. The reason for low stability in water is the autodigestion, which reduces the half-life of the catalyst and makes the product purification difficult. Tüchsen and Ottesen demonstrated that cross-linked subtilisin crystals were resistant to the autodigestion, which rapidly decreased the activity of soluble subtilisin in aqueous solvent.

Wang et al. showed that cross-linked subtilisin crystals retained full activity after 7 d incubation in aqueous buffer at 60 °C, 50% acetone at 40 °C, and in 70% DMF at 25 °C. Crude enzyme extract lost half of its activity after 11 h in aqueous buffer and after 4 d in 50% acetone. Cross-linked subtilisin crystals were highly active in various organic solvents. However, cross-linked crystals required a small amount of water in neat organic solvents, especially in a polar solvent such as acetone or DMF. The total absence of water resulted in the loss of catalyst selectivity and ultimately in the complete loss of activity. The use of the cross-linked crystalline form of subtilisin in neat organic solvents enhances the enzyme activity compared with lyophilized subtilisin precipitate.

The activity of lyophilized subtilisin in nonaqueous media is affected by the pH of the solution, where the lyophilization is conducted (so called "pH memory"). When lyophilized enzymes are suspended in nonaqueous media, the protonation status of these groups does not change. In the case of cross-linked crystals, water can be removed by repeated washing with a water-miscible solvent. The solvent diffuses through the channels in the cross-linked crystals and the interstitial water diffuses out. Since this occurs while there is still water present, the protonation state adjusts accordingly, as the protons are not kinetically trapped.

However, Montaneñez-Clemente et al. showed that the activity of subtilisin Carlsberg was enhanced by lyophilization with methyl-β-cyclodextrin (MβCD). This method yielded a catalyst that was more active than cross-linked crystals of subtilisin. In all demonstrated reactions the MβCD-co-lyophilized enzyme was more active and as enantioselective as the cross-linked crystalline preparation. In the synthesis of peptides and peptidomimetics, the use of polar, water miscible organic solvents is useful since it improves the substrate solubility and shifts the equilibrium towards products.

Acetonitrile containing 1% water was found to be the best solvent for the synthesis. In the same conditions the soluble enzyme was inactivated rapidly. Crosslinked subtilisin crystals accepted both L- and D-amino acid amides as nucleophiles and showed a similar reaction rate for both enantiomers. One major benefit with the crosslinked subtilisin crystals is the lack of autodigestion. A peptide synthesis catalysed by soluble subtilisin leads to product purification problems as a result of the contaminating peptides produced by subtilisin autodigestion.

Crosslinked thermolysin crystals catalysed several peptide syntheses. The specific activities of soluble and cross-linked crystalline thermolysin were very close to each other. Also a 30 amino acid peptide, an oxidised insulin B chain, was a suitable substrate for cross-linked thermolysin crystals with an initial rate of 4% of that observed for a soluble catalyst. However, in the cross-linked crystalline form thermolysin was considerably more stable than the soluble enzyme in ethyl acetate and in mixtures of water-miscible solvents (DMF, THF, acetone) and water.

The catalyst accepted a broad range of substrates and the reaction yields were quantitative. This hydrolysis method renders it possible to carry out the reactions in mild conditions. Normally harsh conditions are needed for regioselective hydrolysis leading to product racemication. Cross-linked subtilisin crystals have also been used for the hydrolysis of proteins, but they suffered from severe mass-transfer limitations of the high molecular weight substrates.

The cross-linked subtilisin catalyst was found to be excellent as a result of the high activity and stability in organic solvents, high regio- and enantioselectivity, and mild reaction conditions. The reusability of cross-linked subtilisin crystals was demonstrated by the resolution of a certain amine. After seven resolution cycles cross-linked crystals retained full activity without loss in enantioselectivity. The crude enzyme extract lost 50% of its activity after two cycles.

#### *Lipases*

Lipases can be used for stereoselective ester conversions, regioselective and chemoselective acylations and deacylations, and simple hydrolysis of esters under mild reaction conditions. Most often crude enzyme preparations have been used as biocatalysts because of their availability and low cost. However, crude lipase mixtures suffer from stability problems in organic solvents and they contain contaminating side activities that have unexpected synthetic properties affecting the enzyme catalysis. Lalonde et al. identified competing hydrolase activities and protease activity in a crude extract of C. rugosa lipase. Cross-linked lipase crystals are considerably more stable in the presence of organic solvents than the soluble crude lipase preparations.

Neither cross-linked soluble lipase nor its precipitate lacking crystalline structure exhibit stability beyond the level of soluble enzyme. Soluble pure lipases cannot be used in synthesis as a result of low stability against high temperatures and organic solvents. The stabilisation effect in organic solvents is extremely important, since many organic compounds are poorly soluble in an aqueous solvent. Both substrate specificity and enantioselectivity of cross-linked lipase crystals are markedly different from those of crude enzyme extract, which shows the benefit of using pure protein catalysts. In the preparation of cross-linked lipase crystals the enzyme is simultaneously purified in the crystallisation.

The use of cross-linked C. rugosa lipase crystals in organic solvents greatly improved enantioselectivities of several reactions including the resolution of acids by esterification, acylation of secondary alcohols with different carboxylic acids and with activated esters, and the ester hydrolysis. The increased enantioselectivity was attributed to the purity of the protein. Persichetti et al. showed that cross-linked lipase crystals were six orders of magnitude more active than lyophilized lipase in the resolution of menthol with vinyl butyrate in toluene.

Lalonde et al. conducted preparative resolutions (100 g scale) of methyl-Ibuprofen and Ketoprofen by cross-linked lipase. The crystalline catalyst was recycled 18 times. During this time the specific activity dropped by 44%. However, immobilised lipase preparation lost virtually all of its activity after the first reaction cycle. Browne et al. studied the production of inherently chiral calixarenes by cross-linked crystals of lipases from Mucor miehei (EC 3.1.1.3) and Aspergillus niger (EC 3.1.1.3). The use of A. niger lipase in a cross-linked crystalline form improved the enantioselectivity of the reaction over chemical catalysis and free C. rugosa lipase catalysed synthesis.

The scale-up of chiral resolution using cross-linked crystalline lipase from Pseudomonas cepacia. The reaction was the chiral resolution of Rsec-phenethyl alcohol from racemic R/S-sec-phenethyl acetate. The reaction parameters were first studied in small-scale (0.2-1 l) in order to determine the optimum conditions for a 100 l reactor. The reaction mixture contained two phases: aqueous phase and organic phase consisting of the substrate R/S-sec-phenethyl acetate. Cross-linked lipase catalyst was not detrimentally affected by a high shear in the reactor needed to obtain homogenous mixing in the two-phase system even in large-scale. In small-scale, the reaction yield was 48% and the product was formed with 99% enantiomeric excess. These values were not changed in the 100 l scale. A total number of nine runs were made with the same batch of cross-linked lipase crystals.

In the first run, 50 g of crystals were added to the reaction mixture (0.5 g $l^{-1}$). After nine runs about 12.5 g of lipase crystals were collected from the reactor. This decrease in the catalyst amount was a result of the problems in the catalyst recovery process after each run. Also the specific activity of the cross-linked catalyst decreased to one third of the initial value. In addition to catalytic applications, the cross-linked crystalline form of lipase can be used to determine the reaction kinetics and the parameters controlling the enantioselectivity in non-aqueous media, which cannot be studied by pure native enzymes as a result of their instability. Cross-linked crystals are the only pure form stable and active enough for mechanistic studies.

#### *Dehydrogenases*

The use of dehydrogenases in synthetic applications is often limited by the instability of both the enzymes and their nicotinamide cofactors. St. Clair et al. crystallised and glutaraldehyde cross-linked horse liver alcohol dehydrogenase (HLADH) in the presence of the reduced nicotinamide adenine dinucleotide (NADH), which was tightly bound to the enzyme by cross-linking. No leakage of cofactor from cross-linked horse liver alcohol dehydrogenase crystallised with NADH was detected in a prolonged incubation of concentrated protein crystal slurry. This was expected, as in the presence of the cofactor, horse liver alcohol dehydrogenase undergoes a large conformational change enclosing the cofactor and, thus, preventing its dissociation from holoenzyme after crystals were cross-linked.

Cross-linked crystals were active catalysts without any cofactor addition and catalysed a reduction of several ketones. When HLADH was crystallised in the absence of NADH, the activity of the resulting cross-linked crystals was low even in the presence of exogenous NADH. The crystallisation and cross-linking of HLADH in the presence of NADH greatly improved the stability of the enzyme and the cofactor. The half-lives of soluble HLADH and NADH were about 14 d and 3 d in a buffer solution at 30°C, respectively. The cross-linked HLADH-NADH crystals maintained full activity after three months of incubation.

Especially the stability of HLADH in the presence of organic solvents is important, because high concentrations of isopropanol, ethanol, and butanediol can be used for cofactor regeneration. The half-life of cross-linked HLADH-NADH crystals in 90% isopropanol was 2 d at 25 °C, for example. Soluble HLADH precipitated and lost activity immediately in these conditions. The cofactor recycle inside cross-linked HLADH-NADH crystals was demonstrated by the reduction of cinnamaldehyde in the presence of 200 mM of 1,4-butanediol to regenerate the cofactor. A column packed with approximately 50 mg of HLADH-NADH containing about 1.2 μmol NADH reduced 15 μmol cinnamaldehyde to cinnamyl alcohol.

Sobolov et al. studied the production of lactate from pyruvate in an electrolytic cell catalysed by cross-linked lactate dehydrogenase crystals. The cross-linked crystalline form of the enzyme was used in the study solely because of the improved stability. Lactate dehydrogenase produces lactate from pyruvate and NADH. Only 10% of the activity of cross-linked crystalline lactate dehydrogenase was lost during 25 days of operation, while soluble lactate dehydrogenase lost its activity totally after 5 days. Furthermore, the cross-linked lactate dehydrogenase crystals could be recycled. The yield of lactate from pyruvate was 70% with cross-linked lactate dehydrogenase crystals. The yield with soluble enzyme was 50%, even though an aliquot containing the same amount of fresh enzyme that was present in the corresponding experiment in the form of crosslinked crystals was added daily to the reaction mixture.

### Other Enzymes

Leisola et.al have studied cross-linked xylose isomerase (CLXIC) and xylanase crystals in packed bed reactors. Because of the high enzyme to substrate ratio, they were able to identify new substrates for xylose isomerase (XI). They used a CLXIC reactor for isomerisations of a variety of pentose and hexose sugars, such as for the production of an unnatural sugar, L-ribose, from L-arabinose. However, the reported efficiency of this conversion was low compared with chemical synthesis or catalysis by a specific enzyme. They further showed that the packed bed reactor was capable of catalysis and simultaneous product separation. The reason for this was that each of the sugars (substrates and products) had a different reaction rate and binding characteristics to the CLXIC matrix.

Similar product separation was observed in the hydrolysis of xylotetraose to xylose by cross-linked xylanase crystals in a packed-bed reactor. Sobolov et al. characterised cross-linked fructose diphosphate aldolase crystals for the synthesis of tagetitoxin. Soluble fructose diphosphate aldolase was an effective synthetic catalyst but suffered from instability, water insolubility of unnatural substrates, and high cost. However, the cross-linked crystalline form was highly stable without any observable loss of activity after 6 months of incubation in room temperature. In the same conditions the soluble enzyme lost its activity after 3-10 days.

The specific activity of cross-linked crystals was 65% of the activity obtained with the soluble enzyme. In organic solvents (e.g., DMSO, acetonitrile, dioxane) the soluble form of fructose diphosphate aldolase lost 75-97% of the initial specific activity after 1 h incubation. However, with cross-linked crystals no significant loss of activity was observed. Hydroxynitrile lyase is used for asymmetric syntheses of intermediates for pharmaceuticals and agrochemicals. Hydroxynitrile lyase have been used both in soluble and carrier-immobilised forms. However, the stability of these enzyme preparations is relatively low both in aqueous and organic media.

Costes et al. prepared cross-linked hydroxynitrile lyase crystals and compared this with immobilised preparates. They used dry acetone in the cross-linking reaction in order to reduce the cross-linking time and glutaraldehyde concentration. They found that by increasing the concentration of glutaraldehyde in cross-linking the resulting crystals were more stable. The specific activity of cross-linked crystals was two orders of magnitude lower than the activity of Celite-immobilised hydroxynitrile lyase.

However, the cross-linked crystals were much more stable in the presence of organic solvents. After seven one-hour reactions in a dibutyl ether solution, only 1% of the initial activity of Celite-immobilised enzyme was left. In contrast, more than 70% of the initial activity of the cross-linked crystals was retained in the same conditions. An organophosphorous hydrolase has been recognised to hydrolyze P-O, P-F, P-S, and P-CN bonds making a variety of chemical agents nonhazardous, including chemical warfare nerve gas VX. However, for practical applications the activity and stability has to be enhanced.

### *Supercritical Fluids*

Supercritical fluids (sc-fluids) are liquids that are heated under pressure until the vapor phase is as dense as the liquid phase. sc-fluids have liquid-like densities and gas-like diffusivities. They are an environmentally friendly alternative to organic solvents for biocatalysis, because the solvent can be completely removed after depressurisation. The most common sc-fluids are sc-$CO_2$ and sc-ethane. Barreiros and co-workers have studied the properties of cross-linked subtilisin crystals in sc-fluids. The model system was cross-linked crystalline subtilisin catalysing transesterification reactions.

Cross-linked crystals in sc-ethane were 2-10-fold more active than in hexane or acetonitrile under similar conditions. The specific activity of cross-linked crystals in scethane was 5-100-fold higher than the specific activity of lyophilized subtilisin. Especially with low water activities cross-linked crystals were superior. The activity of cross-linked subtilisin was further enhanced by addition of salt hydrates or zeolites to the reaction media. This was probably a result of acid-base buffer effect of these additives in low water-activity media. The enzymatic catalysis by cross-linked crystalline enzymes in sc-carbon dioxide has been patented recently.

Protein crystals have many properties that make them interesting separation matrices. They are made of L-amino acids that create an asymmetric environment. The interior of many globular proteins is hydrophobic. Several amino acid residues are either negatively or positively charged leading to ion exchange properties of proteins. Enzymes have also specific binding sites for their substrates and cofactors. In addition, protein crystals are microporous materials. Immobilised proteins have been used in chiral separations. Vilenhick et al. showed for the first time that cross-linked protein crystals can also be used in chiral separations. They described the chiral separations of phenylglycine and phenyllactic acid by cross-linked thermolysin crystals. These compounds are not substrates for thermolysin.

Moreover, they showed that the cross-linked precipitate of human serum albumin did not separate the two enantiomers of folinic acid, while crosslinked human serum albumin crystals gave good chiral resolution. This result suggests that the crystallinity is needed for good separation characteristics. The high surface area of crystalline proteins explained this result. Pastinen et al. used cross-linked xylose isomerase crystals CLXIC in a chiral separation of a racemic mixture of D/L-arabitol. However, arabitol is also a substrate analog of the enzyme and the separation can be a result of the chiral nature of the protein, the specific binding of arabitol to the active site, or a combination of both. CLXIC showed also weak chiral separation ability for some D/L pairs of some amino acids.

CLXIC separated n-alcohols from $C_1$ to $C_8$ based on the hydrophobic interaction between the carbon chain of the alcohols and the protein. The retention time of 1-octanol was approximately doubled compared with the retention time of methanol. CLXIC separated L-forms of tryptophane, histidine, tyrosine, and phenylalanine from other natural amino acids. Vilenchik et al. showed that cross-linked thermolysin crystals separated a mixture of

Ibuprofen and phenyllactic acid. The mechanism of these separations is not known. Pastinen et al. used CLXIC as an affinity separation matrix for sugar alcohols xylitol and sorbitol, which are substrate analogs to xylose isomerase.

Sugar alcohols were specifically bound to CLXIC even from impure solutions, and were eluted by dilute $CaCl_2$. Only the active form of the enzyme had the separation ability. However, only one molecule of xylitol or sorbitol was bound to each active center of xylose isomerase, which led to a low overall separation capacity of 1 g xylitol per 500 g CLXIC. Both Pastinen et al. and Vilenchik et al. pointed out the good mechanical stability of a cross-linked protein crystal stationary phase. A cross-linked xylose isomerase column was packed with 30 MPa pressure, and thermolysin and human serum albumin columns with 10 MPa pressure, which shows the rigidity of the crystals.

## X-ray Crystallography

All protein crystals are sensitive to radiation used in X-ray crystallography, which is an essential tool in the structure determination of proteins. In some cases the crystal sensitivity is unmanageable and the diffraction quality of crystals deteriorates rapidly. Quiocho and Richards initially described the concept of cross-linked protein crystals in 1964. They stabilised carboxypeptidase A crystals by glutaraldehyde treatment for Xray structure analysis. There are only minor changes in the diffraction pattern of crosslinked protein crystals compared with that of the native crystals.

Crosslinked crystals render it possible to study the protein structure in organic solvents, which cannot be done with native crystals. In cryocrystallography, crystals are treated with a cryoprotectant (e.g., sugar solutions, ethylene glycol, PEG, glycerol) and cooled rapidly to cryogenic temperatures (e.g., with nitrogen gas around 100

K). However, the crystals often suffer damage on cooling. Lusty showed that glutaraldehyde cross-linking protected the crystals from the damage on cooling without any effect on the diffraction properties.

### Design of New Protein Materials

Dotan et al. described the production of a pre-designed protein crystal lattice. They chose a nearly tetrahedral lectin concavalin A as a model building block. This lectin is a tetramer with a fully characterised three-dimensional structure. The lectin tetramer contains four binding sites for α-D-mannopyranoside or α-D-glucopyranoside. They used computer modeling, which suggested that the cross-linking of soluble lectin tetramers by bismannoside would result in the formation of a diamond-like three-dimensional protein lattice. Indeed, the addition of bismannopyranoside to the lectin solution resulted in a rapid formation of lectin crystals, which were simultaneously cross-linked. A chemical analysis of the bismannopyranoside content in the crystals supported the hypothesis that each lectin molecule was cross-linked through four molecules to its neighbors.

The size of the resulting crystals was 0.1-100 µm depending on the pH in the crystallisation. The cross-linked crystalline product was insoluble and resistant to pH changes. Häring and Shreier combined the methods of chemical modification of the enzymatic framework as well as the enzymatic active site. First they prepared crosslinked crystals of protease subtilisin. Then they activated the active site serine 221 by PMSF and subsequently PMSF was substituted by sodium hydrogen selenide. The selenol group was oxidised with hydrogen peroxide to yield a stable seleninic acid form of cross-linked subtilisin crystals. As a result, the semisynthetic seleno-subtilisin showed glutathione peroxidase activity, and catalysed kinetic resolution of several racemic hydroperoxides. The preparation of seleno-subtilisin has been described earlier. However, the crystalline seleno-subtilisin showed a high specific activity, and stability against organic solvents and high temperatures compared to amorphous seleno-subtilisin.

### *Applications of Cross-linked Protein Crystals*

St. Clair et al. investigated the possibility to use cross-linked protein crystals as antigens. Currently the progress is toward subunit vaccines, which are composed of a purified antigenic determinant separated from the disease-causing organism. To enhance the immune response, subunit vaccines universally require adjuvants and delivery vehicles. Human serum albumin in a cross-linked crystalline form showed up to 30-fold increased immunogenity and increased stability over the soluble albumin that was formulated with a commercial aluminium adjuvant.

Additionally, Margolin et al. crystallised and carbohydrate cross-linked several vaccine antigens to formulate vaccines. However, no details of the function of these antigens as vaccines were provided. Margolin and Navia described in their review that cross-linked protein crystals could be used in oral lumenal therapy. Metabolic or gastrointestinal diseases are often the result of impaired or missing enzymes or protein function in the biochemical pathway. However, the conditions in the gastrointestinal track are challenging for protein drug delivery.

The pH is acidic ($pH < 2$) and there is a high concentration of proteases in the stomach and gut. Because of the excellent stability and good activity of proteins in the cross-linked crystalline form, the patient can orally administer a cross-linked crystalline therapeutic protein to perform the desired reactions in the gut lumen. According to Margolin and Navia preliminary results indicate that in a cross-linked crystalline form lipases are at least 100-fold more active and significantly more stable than commercial lipase products. These principles are also described in a recent patent application.

### *Use of Biosensor*

De Mattos et al. compared glucose biosensors prepared from commercial lyophilized glucose oxidase, crystalline glucose oxidase, and cross-linked crystalline glucose oxidase. The best performance in terms of sensitivity, detection limit, and operational stability was obtained with biosensors based on cross-linked protein crystal technology. Probably, the enzyme stability during immobilisation, which includes drying the enzyme layer and contact between dried glucose oxidase and ethanol, is critical. The cross-linked crystals seemed more resistant to denaturation upon the preparation of the sensor.

# 6

# Membrane Separation Processes

The membrane can be defined essentially as a barrier, which separates two phases and restricts transport of various chemicals in a selective manner. A membrane can be homogenous or heterogeneous, symmetric or asymmetric in structure, solid or liquid, can carry a positive or negative charge or be neutral or bipolar. Transport through a membrane can be effected by convection or by diffusion of individual molecules, induced by an electric field or concentration, pressure or temperature gradient. The membrane thickness may vary from as small as 100 rnicron to several mms.

A membrane separation system separates an influent stream into two effluent streams known as the permeate and the concentrate. The permeate is the portion of the fluid that has passed through the semi-permeable membrane. Whereas the concentrate stream contains the constituents that have been rejected by the membrane. Membrane separation process enjoys numerous industrial applications with the following advantages:

— Appreciable energy savings

— Environmentally benign

— Clean technology with operational ease

— Replaces the conventional processes like filtration, distillation, ion-exchange and chemical treatment systems

— Produces high,quality products

— Greater flexibility in designing systems.

**Different Types of Membranes**

The proper choice of a membrane should he determined by the specific application objective: particulate or dissolved solids removal, hardness reduction or ultra pure water production, removal of specific gases/chemicals etc. 1'he end-use may also dictate selection of membranes for industries such as potable water, effluent treatment, desalination or water supply for electronics or pharmaceutical manufacturing. Following are the types of membranes commonly used.

— *Microporous Membranes*: The membrane behaves almost like a fibre filter and separates by a sieving mechanism determined by the pore diameter and particle size. Materials such as ceramics, graphite, metal oxides, polymers etc. are used in making such membranes. The pores in the membrane may vary between 1 nm-20 microns.

—*Homogeneous Membranes*: This is a dense film through which a mixture of molecules is transported by pressure, concentration or electrical potential gradient. Using these membranes, chemical species of similar size and diffusivity can be separated efficiently when their concentrations differ significantly.

— *Asymmetric Membranes*: An asymmetric membrane comprises a very thin (0.1-1.0 micron) skin layer on a highly porous (100-200 microns) thick substructure. The thin skin acts as the selective membrane. Its separation characteristics are determined by the nature of membrane material or pore size, and the mass transport rate is determined mainly by the skin thickness. Porous sub-layer acts as a support for the thin, fragile skin and has little effect on the separation characteristics.

— *Electrically Charged Membranes*: These are necessarily ion-exchange membranes consisting of highly swollen gels carrying fixed positive or negative charges. These are mainly used in the electro-dialysis.

— *Liquid Membranes*: A liquid membrane utilises a carrier to selectively transport components such as metal ions at relatively high rate across the membrane interface.

**Membrane Modules**

The membranes can he cast as flat sheets, tubes and fine hollow fibres. For accommodating such shapes and structures, different types of membrane modules are available. The last decade of membrane and module development has lessened the effects of physical compaction and has brought forth spiral membrane modules capable of operating at pressures in excess of 800 psig (55.2 bar). The techno-economic factors for the selection, design and operation of membrane modules include cost of supporting materials and enclosure (pressure vessels), power consumption in pumping and ease of replaceability.

The following membrane modules are largely used for industrial applications:

— Plate and frame module

— Spiral wound module

— Tubular membrane module

— Capillary membrane module

— Hollow fibre membrane module.

### Types of Membrane Separation

Various types of membrane separation have been developed for specific industrial applications. Some of the widely used processes are described here:

#### *Reverse Osmosis (RO)*

Unlike water filtration, that can only remove some suspended materials larger than 1 micron, the process of reverse osmosis (RO) will eliminate the dissolved solids, bacteria, viruses and other germs contained in the water. RO is essentially a pressure driven membrane diffusion process for separating dissolved solutes. The RO is generally used for desalination seawater for its conversion into potable water. The salient features of the process are that it involves no phase change and it is relatively a low energy process. Almost all RO membranes are made polymers, cellulosic acetate and matic polyamide types rated at 96%-99+% NaCl rejection.

RO membranes are generally of two types, asymmetric or skinned membranes and thin film composite (TFC) membranes. The support material is commonly polysulfones while the thin film is made from various types of polyamines, polyureas etc. RO membranes have the smallest pore structure, with pore diameter ranging from approximately 5-15 A0 (0.5 nm - 1.5 nm). The extremely small size of RO pores allows only the smallest organic molecules and unchanged solutes to pass through the semi-permeable membrane along with the water. Greater than 95-99% of inorganic salts and charged organics will also be rejected by the membrane due to charge repulsion established at the membrane surface. RO finds extensive applications in the following:

— Potable water from sea or brackish water

— Ultrapure water for food processing and electronic industries

— Pharmaceutical grade water

— Water for chemical, pulp & paper industry

— Waste treatment etc.

*Future directions*

— Municipal and industrial waste treatment

— Process water for boilers

— De-watering of feedstreams

— Processing high-temperature feed- streams etc.

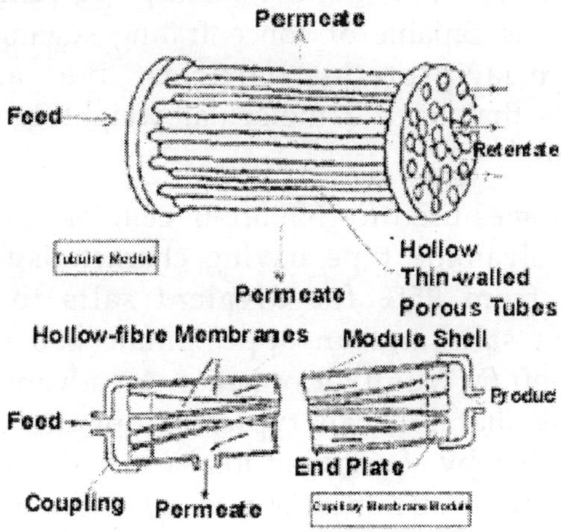


*Figure 1. Tubular membrane module and capillary membrane module*

In the last six to eight years the technology has gained industry acceptance as a viable water treatment option for many different fluid separation applications. Low operating costs and the ability to remove organic contaminants and 95-99% of inorganic salts with minimal chemical requirements make RO an attractive technology for many industrial applications. A complete RO equipment comprising pre-treatment system, membrane modules etc. is estimated to cost around US $ 450 per m3 of flow per day.

### *Nanofiltration (NF)*

Nanofiltration is a form of filtration that uses membranes to separate different fluids or ions. NF is typically referred to as "loose" RO due to its larger membrane pore structure as compared to the membranes used in RO, and allows more salt passage through the membrane. Because it can operate at much lower pressures, and passes some of the inorganic salts, NF is used in applications where high organic removal and moderate inorganic removals are desired. NF is capable of concentrating sugars, divalent salts, bacteria, proteins, particles, dyes and other constituents that have a molecular weight greater than 1000 daltons.

Membranes used for NF are of cellulosic acetate and aromatic polyamide type having characteristics as salt rejections from 95% for divalent salts to 40% for monovalent salts and an approximate 300 molecular weight cut-off (MWCO) for organics. An advantage of NF over RO is that NF can typically operate at higher recoveries, thereby conserving total water usage due to a lower concentrate stream flow rate. NF is not effective on small molecular weight organics, such as methanol.

### *Ultrafiltration (UF)*

Ultrafiltration is most commonly used to separate a solution that has a mixture of some desirable components and some that are not desirable. UF is somewhat dependent on charge of the particle, and is much more concerned with the size of the particle. Typical rejected species include sugars, bio-molecules, polymers and colloidal particles. The driving force for transport across the membrane is a pressure differential. UF processes operate at 2-10 bars though in some cases up to 25-30 bars has been used. UF processes perform feed clarification, concentration of rejected solutes and fractionation of solutes.

UF is typically not effective at separating organic streams. UF membranes are capable of retaining species in the range of 300-500,000 daltons of molecular weight, with pore sizes ranging from 10-1000 Angstroms (103-0.1 microns). These are mostly described by their nominal molecular weight cutoff (1000-100,000 MWCO), which means, the smallest molecular weight species for which the membranes have more than 90% rejection. UF usually implies separation of macromolecules such as protein from low molecular weight solvents.

Pores in the support layer of the membrane are relatively larger than those of the surface layer. Material passing into fine pores can readily be transported through the open-celled, sponge-like structure of the support layer. For example, in electrodeposition paint recovery, the paint, composed of resin, a pigment and water are separated into two streams that can he reused. The first stream includes the water and a small amount of paint resin, which can he used to rinse the parts later in the process. The paint pigment is separated from that stream and can be re-used in the paint bath, allowing the bath to be concentrated to a useable level.

It is found that, whenever the solvent of a mixture flows through the membrane, retained species are locally concentrated at the membrane surface, thereby resisting the flow. In the case of processing solution, this localised concentration of solute normally results in precipitation of a solute gel over the membrane. When processing a suspension, the solids collect as a porous layer over the membrane surface.

The permeate rate can be effectively controlled by the rate of transport through the polarisation layer rather than by membrane properties. Hence, UF throughput depends on physical properties of the membrane, such as permeability, thickness, process and system variables like feed consumption, feed concentration, system pressure, velocity and temperature. UF has a wide range of applications as shown below:

— Oil emulsion waste treatment

— Treatment of whey in dairy industries

— Concentration of biological macromolecules

— Electrocoat paint recovery

— Concentration of textile sizing

— Concentration of heat sensitive proteins for food additives

— Concentration of gelatin

— Enzyme & pharmaceutical preparations

— Pulp mill waste treatment

— Production of ultrapure water for electronics industry

— Macromolecular separations replacing the conventional change of phase methods.

The important characteristics for membrane materials are porosity, morphology, surface properties, mechanical strength and chemical resistance. Polymeric materials, viz., polysulfone, polypropylene, nylon 6, Polytetrafluoro-ethylene (PTFE), PVC, acrylic copolymer etc. have been used successfully as UF membranes. Inorganic materials such as ceramics, carbon based membranes, zirconia etc. have been commercialised by several vendors. UF may find wide range of applications in the near future and some of those processes important from the separation and energy savings point of view are mentioned below:

### Microfiltration (MF)

This is by far the most widely used membrane process with total sales greater than the combined sales of all other membrane processes. Microfiltration has numerous small applications. It is essentially a sterile filtration with pores (0.1-10.0 microns) so small that micro-organisms cannot pass through them. Microfiltration is a process of separating material of colloidal size and larger than true solutions. A MF membrane is generally porous enough to pass molecules of true solutions, even if they are large.

Microfilters can also he used to sterilise solutions, as they are prepared with pores smaller than 0.3 microns, the diameter of the smallest bacterium, pseudomonas diminuta. While the mechanism for conventional depth filtration is mainly adsorption and entrapment, MF membranes uses sieving mechanism with distinct pore sizes for retaining larger size particles than the pore diameter. Hence, this technology offers membranes with absolute rating, which is highly desirable for critical operations such as sterile filtration of parental fluids, sterile filtration of air and preparation of particulate,free-water for the electronics industry.

The MF membranes are made from natural or synthetic polymers such as cellulose nitrate or acetate, polyvinylidene difluoride (PVDF), polyamides, polysulfone, polycarbonate, polypropylene, PTFE etc. The inorganic materials such as metal oxides (alumina), glass, zirconia coated carbon etc. are also used for manufacturing the MF membranes. The properties of membrane materials are directly reflected in their end applications. Some criteria for their selection are mechanical strength, temperature resistance, chemical compatibility, hydrophobility, hydrophilicity, permeability permselectivity and the cost of membrane material as well as manufacturing process. MF has a wide array of applications as mentioned below:

— Preparation of parenterals and sterile water for pharmaceutical industry

— Food & beverages (concentration of fruit juices and alcoholic beverages

— Chemical industry

— Microelectronics industry

— Fermentation

— Laboratory/analytical uses etc.

The following are the likely applications of MF in the near future:

— In biotechnology for concentration of biomass, separations of soluble products

— In diatomaceous earth displacement

— In non-sewage waste treatment for removing intractable particles in oily fluids, aqueous wastes which contain particulate toxics and stack gas

— In paints for separating solvents from pigments etc.

***Electrodialysis (ED)***

This is an electro-membrane process in which the ions are transported through a membrane from one solution to another under the influence of an electrical potential. ED can be utilised to perform several general types of separations such as separation and concentration of salts, acids and bases from aqueous solutions or the separation and concentration of monovalent ions from multiple charged components or the separation of ionic compounds from uncharged molecules. ED membranes are usually made of cross-linked polystyrene that has been sulfonated. Anion membranes can he of cross-linked polystyrene containing quaternary ammoniuri groups.

Usually, ED membranes arx fabricated as flat sheets containing about 30-50% water. Membranes are fabricated by applying the cation and anion-selective polymer to a fabric material. The system consists of two kinds of membranes: cation and anion, which are placed in an electric field. The cation-selective membrane permits only the cations, and anion-selective membrane only the anions. The transport of ions across the membranes results in ion depiction in some cells, and ion concentration in alternate ones. Electrodialysis is used widely for production of potable water from sea or brackish water, electroplating rinsc recovery, desalting of cheese whey, production of ultrapure water etc.

The present ED industry has experienced a steady growth rate of about 15% since 15 years. To ensure further growth beyond desalination and salt production, new areas of application are being exploited in the areas of chemicals and pharmaceuticals, food, industrial & municipal effluent treatment etc. This can be achieved through slight modifications in the conventional processes as well as extensive R&D work in those areas. The following are some new business opportunities in ED separation process:

- De-ionized water from conductive spacers
- Radioactive wastewater treatment by using radiation resistant membranes
- Deacidification of fruit juices
- Heavy metal recovery
- Recovery of organic acids from salts

— pH control without adding acid or base

— Regeneration of ion-exchange resins with improved process design

— Acid recovery from etching baths etc.

### *Gas Separation*

The membrane gas separation technology is over ten years old and is proving to he one of the most significant unit operations. These processes compete with technology alternatives such as adsorption, cryogenic distillation etc. in niche application areas. The membrane processes enjoy certain advantages, viz., compactness and light in weight, low labour intensity, modular design permitting easy expansion or operation at partial capacity, low maintenance (no moving parts), low energy requirements and low cost especially so for small sizes.

Membranes made of polymers and copolymers in the forms of flat film or hollow fibre have been used for gas separation. Different gases pass through certain membranes at significantly different rates, thus permitting a partial separation. The rate of permeation is proportional to the pressure differential across die membrane and inversely proportional to the membrane thickness. The rate of permeation is also proportional to the solubility of the gas in the membrane and also to the diffusivity of gas through the membrane.

Gas separation is thus affected by three key performance attributes of membranes, viz., selectivity towards the gases separated, membrane flux or permeability and the life of the membrane, maintenance and replacement costs. The membrane gas separation has been used for hydrogen separation and recovery, ammonia purge gas, refinery hydrogen recovery, 'syngas' separation in petrochemicals industry, $CO_2$ enhanced oil recovery, natural gas processing, landfill gas upgrading, air separation, nitrogen production, air dehydration, helium recovery etc. The gas separation technology may enjoy the following applications in the near future:

— $N_2$ enrichment of air

— Low level $O_2$ enrichment of air

— H2 and acid gas separation from hydrocarbons

— Helium recovery

— Natural gas dehydration

#### *Pervaporation*

This is a membrane based process for separating miscible liquids. Here the absorption of one of the components of the liquid by the membrane, diffusion of this component across the membrane and evaporation, as permeate vapour, into the partial vacuum applied to the underside of the membrane. Pervaporation differs from all other membrane processes because of the phase change of the permeate, non-porous. Transport through these membranes is effected by maintaining a vapour pressure gradient across the membrane:

Applications of pervaporation membranes: Pervaporation offers significant capital and energy savings in applications that are difficult to separate by more conventional techniques such as azeotropic mixtures or mixtures of close boiling components. Pervaporation has been used for separation of ethanol-water mixture, solvent recovery, separation of heat sensitive products or enrichment of organic pollutants etc. Advantages of this process over other separation techniques are given below:

— Effective and economic separation of mixtures of substances with small difference in boiling point and azeotropic mixtures

*Table 1. characteristics of membranes used in different membrane separation processes*

| Process | Membrane Type and Pore Radius | Membrane Material | Process Diving Force | Applications |
|---|---|---|---|---|
| Microfiltration | Symmetric microporous, 0.1-10 microns | Cellulose nitrate or acetate, Polyvinylidene difluoride (PVDF), Polyamides, Polysulfone, PTFE, Metal Oxides etc. | Hydro-static pressure difference at approx. 10-500 kPa | Sterile filtration, Clarification |
| Ultrafiltration | Asymmetric microporous, 1-10 nm | Polysulfone, Polypropylene, Nylon 6, PTFE, PVC, Acrylic Copolymer | Hydrostatic pressure difference at approx. 0.1-1.0 Mpa | Separation of macromole-cular solutions |
| Reverse Osmosis | Asymmetric skin-type, 0.5-1.5 nm | Polymers, Cellulosic acetate, Aromatic Polyamide | Hydrostatic pressure difference at approx. 2-10 Mpa | Separation of salts and microsolutes from solutions |
| Electrodialysis | Cation and anion exchange membrane | Sulfonated cross-linked polystyrene | Electrical potential gradient | Desalting of ionic solutions |
| Gas Separation | Asymmetric homogeneous polymer | Polymers & copolymers | Hydrostatic pressure and concentration gradients | Separation of gas mixtures |

| | | | | |
|---|---|---|---|---|
| Pervaporation | Asymmetric homogenous polymer (A non-porous membrane | Polyacrylonitrile, Polymers | Vapour pressure gradient | Separation of azeotropic mixtures |
| Nanofiltration | Thin-film membranes | Cellulosic Acetate and Aromatic Polyamide | 9.3-15.9 bar | Removal of hardness and desalting |

— Modular membrane design
— No entrainers for separation of azeotropic mixtures
— Reduced capital costs compared to conventional systems.

Table 1 represents the characteristics of membranes used in different membrane separation processes, process driving forces and applications of such processes

**Emerging Membrane Technologies**

Membrane based solvent extraction devices, recently developed, appear to eliminate high capital, operating and maintenance costs of centrifugal devices and additionally provide very high volumetric transfer rates. Systems have been designed by using microporous membranes (hydrophobic or hydrophilic) for non-dispersive solvent extraction.

Hollow-fibre-contained-liquid-membranes (HFCLM) have been used for gas separation through a nonporous polymeric membrane. Microporous polypropylene hollow fibres have been used as the membrane material. Gas separations such as $N_2$-$CO_2$, $CH_4$-$CO_2$, $SO_2$-$CO_2$-$N_2$ and others have been studied by HFCLM technique.

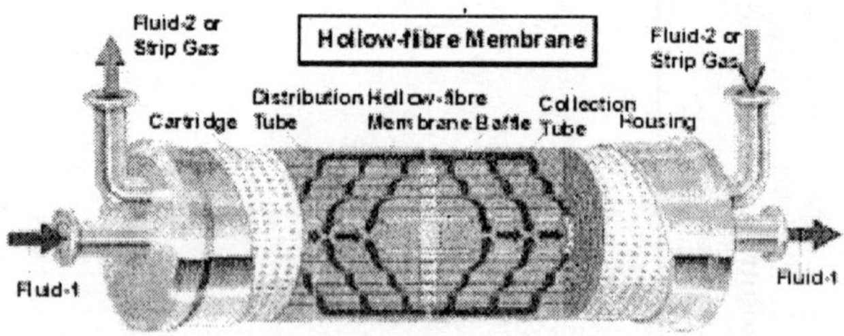


*Figure 2. Hollow-fibre membrane system.*

### Ion-exchange Membranes

A new type of ion-exchange membrane developed by the membrane research group at McMaster University, Canada, is based on the simple concept of taking a cheap, chemically and physically rugged microporous membrane and filling the pores with a flexible, ionic 'jelly'. This gel fill is anchored into place and confined by the microporous substrate. These membranes, which in fact contain 70% water, show remarkable separation properties including very large chemical value effects. One feature of this new class of membranes is that they show 'intelligence' for being able to sense the nature of the solution in contact with the membranes and modify their properties.

These membranes find extensive application in electrodialysis, diffusion dialysis, nanofiltration, membrane solvent extraction and facilitated transport applications. The prototype membranes have excited the interest of a series of companies in different parts of the globe. An interfacial polymerisation/coating technique has been recently developed at McMaster University. It evenly coats all the interior surfaces of a microfiltration membrane with polymers such as polyamides, polyesters, and polysulphonamides.

The coatings which cannot he removed, modify the surface chemistry and provide different types of properties for the membranes. Thus, the membrane remains microporous, but, with different surface chemistry. Photochemically active groups have also been incorporated into the coated layer to modify their chemical properties. Pervaporation membranes are being developed for the removal of trace organics from water using coatings derived from silicone based oligomers.

### Membranes for Fuel Cells/Electrolysis

One breakthrough in the application of ionic conducting

polymer membranes is the proton exchange membrane fuel cell, a device that converts chemical energy directly into electrical energy without burning. As the electrochemical combination of hydrogen (the fuel) and oxygen produce water, the fuel cell is environmentally clean and is expected to replace the gasoline engine or rechargeable battery in the automobiles.

### Encapsulating Membranes

Research in the area of design and preparation of encapsulating media for environmentally sensitive materials addresses phospholipid liposomes modified with synthetic polymers anchored on the membrane outer surface, and non-phospholipid liposomes (NPL) prepared from commercially available surfactants. The aim of the above is to tailor the encapsulation medium and process to specific applications in industrially relevant areas. The liposomes structure and properties are studied by spectroscopic techniques, such as fluorescence, size exclusion chromatography and capillary electrophoresis.

G-50 Ultrafiltration systems replaces the conventional oily wastewater treatment systems which are limited to sludge generation, insufficient removal of oil, and high operating costs. The G-50 UF systems are used to treat oily waste streams to recover water and oil or to meet discharge regulations. G,50 UF systems generally designed to operate in batch mode at a net pressure of 620-825 kPa. Advantages of G-50 UF systems are;

- G-50 UF systems are more compact and require less floor space than conventional oil removal equipment
- Coagulants and filter aids are not required
- Oil can be concentrated for reuse
- Cost effective.

## Supported Liquid Membrane (SliM) Process

Commodore's SliM is the first membrane technology that is capable, in a single process, of selectively extracting multiple elements or compounds ftorn a mixed process stream. The conventional methods for separating metals and/or removal from solubilised process streams presently include ion exchange, RO, UF, nanofiltration, precipitation and chromatography. Most of these methods have certain drawbacks, including lack of selectivity in the separation process, inability to handle certain metals in the process streams and the creation of sludges and other harmful by-products which require further post-treatment prior to disposal. SliM is a superior and cost-effective alternative to the existing forms of membrane filtration technologies, as it has the following advantages:

— Requires lower initial capital costs and lower operating costs

— Capability of treating variety of elements and compounds in industrial settings at greater speed and with a high degree of effectiveness, with varying contaminant concentrations and volume requirements

— Environmentally safe technology

— Total recovery is possible

— No sludge production

— Meets EPA Common Sense Initiative (CSI) guidelines and zero discharge regulations for effluent discharge

— No post-treatment costs

— Improves process quality

The SliM process involves passing a contaminated aqueous or gaseous feedstream through a hollow, porous fibre membrane. This membrane is previously loaded with chemicals whose composition varies depending on the targeted substance in the feedstream. As the feedstream enters the module, the metal or other substance to be extracted reacts with the proprietary chemical combination in the module, and the metallic or other ions are extracted through the membrane into a strip solution which is concentrated and collected in a separate storage container. The balance of the feedstream is either recycled or simply discharged as normal effluent. This is an ideal system for finishers, electroplating shops, PCB manufacturers, remediation sites, landfills and mines.

### Liquid Membrane Extraction

Liquid membranes can be used to extract trace amounts of heavy metal ions frorn process streams in one process step. The liquid membrane is a thin layer of organic phase, separating two aqueous phases. To achieve metal ion extraction, an extractant is dissolved in the organic membrane phase. It acts as a shuttle, extracting the metal ion from the feed phase and releasing it again at the other side of the membrane.

Advantages of the process are the high driving force and the low organic inventory. The second type of liquid membrane is emulsion liquid membrane (ELM), which is water in oil emulsion, stirred in the third water phase. This system is studied by GRM Breembroek. The main subject of the research is the investigation of the influence of various process parameters on the extraction rate in various liquid membrane systems, and the adaption of the system for the treatment of solid-containing process streams.

#### *Present Market Position*

Membrane processes are used widely for desalination, water purification, medical applications, chemical and food processing, but their use as an alternative to distillation has generally reached only the demonstration stage. There are, however, substantial potential markets for membrane applications worldwide which EU manufacturers can exploit, eg in separation of value-added products, gas separation, separation of alkane-alkene mixtures, hydrogen recovery and waste management.

There is a wide range of membrane technologies and most are being developed or marketed by companies in the European Union. Development is at various stages from basic R&D to commercial status. It is expected that most of these technologies will be fully deployed by the year 2010 both inside and outside of the European Union. The potential deployment on the continent is large with the bulk of the sales expected to come from Member State companies. However the world markets should be at least 3 times bigger than the European market and industries will be strongly poised to win a large share of this. The main developers of the technology are the USA and Japan, but European companies have a major potential role in marketing and selling the technology.

The greatest uptake of membrane and crystallisation techniques is expected in applications where they have a natural advantage, eg in the separation of heat-sensitive materials or for azeotropic mixtures.In the food and beverage sectors, membranes are used to concentrate, fractionate and purify liquid products. Four types of membrane processes are important, being microfiltration (MF), ultrafiltration (UF), nanofiltration (NF) and reverse osmosis (RO). Gas separation is only used in the fruit- and vegetable sector for packaging in a nitrogen atmosphere.

The differences between the membrane methods used are the separation capabilities due to size and molecular weight, although the separation capabilities do overlap. To give an idea of the possibilities, an overview of the different food and beverage sectors is given, together with some potential applications. Due to the diverse activities in the food industry, quantitative estimates of the potential capacities for membrane applications per sector are difficult to give.

In the chemical industry, currently the most important of processes are gas separation and microfiltration. These two processes show large potential energy-savings before 2010. Other processes are at the demonstration stage on a limited scale. Since the applications of membranes are very diverse, it is impossible to give general figures regarding the energy and economic parameters. Only rough estimates are available. Despite the rough estimates, generally the energy consumption of membrane processes will be much lower than of conventional separation methods.

### Alkane-Alkylene Separation

The conventional process for alkane-alkene separation is distillation. Around 1996/7, micro-filtration membranes were introduced in this separation. The use of micro-filtration membranes in the alkane-alkylene separation is almost in the commercial phase, however chemical companies still do not consider it to be proven technology. Therefore, the expectations are that the real application of membrane technology substitution will start in sectors that produce high value added products (e.g. pharmaceuticals). A relatively new method is using facilitated transport membranes.

The basic principle of the separation using these membranes is that the molecules of interest are absorbed at the membrane, forming a complex. This complex migrates through the membrane, thus separating the mixture. Smit compare the separation with membranes relative to the conventional cryogenic method. The membrane that is examined is the so-called facilitated transport membrane, a membrane combining high selectivity with high permeability. This membrane has been used by BP for complex and removing olefins from various feed streams.

The typical size of this kind of membrane contactors is about 300 - 1500 $m^2$. BP has been using facilitated transport membranes in two different demonstration projects. These projects were considered successful, regarding the purity of the obtained products. Also calculations have been performed on the economics of the process of ethylene recovery from ethylene oxide plant.

**Hydrogen Recovery**

Hydrogen recovery is important in many (petro-)chemical processes. According to, hydrogen recovery by use of membranes can be used for ammonia purge gas, refineries and in the syngas production. Shaver mentions also naphtha crackers where membranes could be used for hydrogen recovery. Ammonia is typically produced by the reaction of hydrogen and nitrogen at high pressure and temperature over a catalyst. Hydrogen feedstock is produced by steam reforming of natural gas. Because the conversion rate in $NH_3$ synthesis is limited to 18-20% unreacted gases are recycled back to the feed to improve yields.

To prevent build up of inert gases and unreacted methane, gas is purged off. This purge gas contains valuable hydrogen, but separation methods were considered too expensive. The purge gas is usually used as a fuel for the reformers. Membranes were demonstrated to provide a cost-effective separation that results in energy-savings and capacity increase. A schematic diagram of the flow for the single-stage membrane process is shown in the next page.

Hydrogen is recovered at relatively low pressures and has to be recompressed for the process. It can be concluded that membranes are likely to be very attractive competitors if the feed is available at high pressure and the permeate can be tolerated at lower pressure. Schendel compared the economic figures for a membrane based system with cryogenic separation. They found that the investment costs were comparable for both options ($1.35 million, for 450.000m³/day separation), as well as were the hydrogen recovery rates. O&M costs were found lower in the membrane plant.

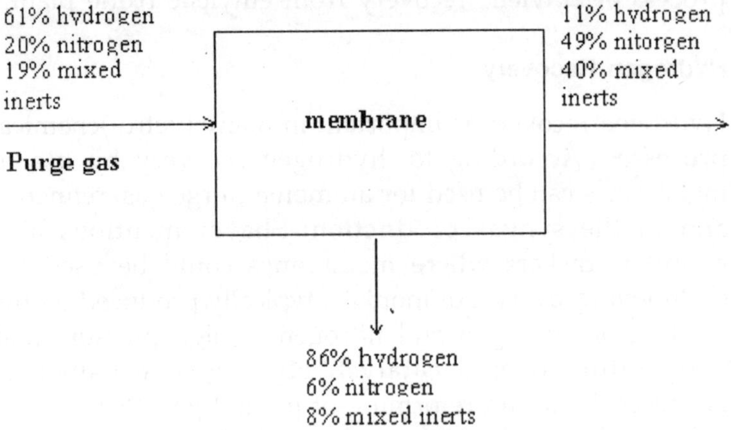


*Figure 3. Material flows in a single stage membrane.*

Even though this comparison is from 1983, the advantages of membranes were clear. Since then, the membrane capital prices have dropped. Lower investment costs for a membrane based system have recently been acknowledged by Koros and Meindersma. Energy savings can be calculated by comparing the hydrogen as a feedstock to hydrogen as a fuel. According to Meindersma and Mulder the energy savings are in the same order as for alkane-alkylene separation; about 5-10% if membranes are used for recovering hydrogen instead of using hydrogen as fuel for the reformers, and in the future this percentage will probably increase.

# 7

# Separation and Production of Bioactive Molecules

Biological recognition-based affinity chromatography was first described by Cuatrecasas and co-work-ers in the late 60s. Since this period, many reviews dealt with this subject: basic procedures for preparation and operation of affinity supports may be found in for example. The most widely used matrixes are agarose derivatives.

These beads are rather sensitive to aggressive physical (particularly pressure and temperature) or chemical environments. The pseudobiospecific ligand affinity chromatography where inexpensive and robust substitutes (metal, dyes etc.) replace biospecific ligands. The design of polymerised imprinted molecules is reviewed in; in this case, no ligand of biological origin is present on the support. Regarding the matrix, the grafting of ligands onto pressure-resistant silica supports was reported by the group of Mosbach.

After the steric recognition between an antigen and its corresponding antibody, close intermolecular bonds stabilise the complex. These additional bonds are those currently occurring in inter- or intramolecular interaction, leading to binding energies ranging from 40 to 70 kJ $mol^{-1}$. The implementation of these bonds is accompanied by extrusion of ordered water (hydration water) from the complex. The antigen — antibody complex formation is associated with a volume increase ranging from 150 to 800 ml $mol^{-1}$, in the case of polyclonal antibodies.

When haptens, such as the fluorescein molecule and various monoclonal antibodies form complexes, the volume change ranges from –12 to +27 ml $mol^{-1}$. This is reminiscent of the variation of volume observed during dimer formation (50-150 ml $mol^{-1}$, reported by Silva and Weber). These different behaviours might have some consequences on the development of baroprocesses. The thermodynamic and the kinetic aspects of this reaction have also been reviewed.

## Disruption of Affinity Complexes

Many strategies have been proposed for antigen elution from immunoaffinity supports, such as extremes of pH, chaotropic salts, ionic strength and denaturants. All these conditions lead to a rapid and irreversible loss of the properties of the immunoadsorbent. For example, a critical study of the long-term degradation mechanisms of six monoclonal antibodies grafted on a Sepharose 4B matrix by various chemical agents was achieved.

In order to act on hydrophobic interactions some authors suggested the use of organic solvents (ethylene glycol, dioxane) combined with temperature control in a DNP-BSA anti-DNP model. Nevertheless, in all cases the eluting reagent has to be further eliminated. Physical methods have then been applied to overcome this drawback. The first application of electrophoresis to achieve antigen-antibody dissociation was reported by Grenot and Cuilleron for the isolation of antibodies to steroid hormones from antigen-linked Sepharose. The major disadvantage was the slow release of antibodies and their high dilution.

To remedy this point, isoelectrophoretic electrodesorption of monoclonal antibodies to transferrin linked to transferrin immobilised on polyethylene glycol tetraphthalate powder was performed by Kasicka and Prusík. This method was developed on a nanoscale. Last, a review relative to principles, methodology, and applications of electrophoretic elution from biospecific adsorbents may be found in.

### Baro-immunodesorption

A physical parameter susceptible to induce dissociation of affinity complexes is the application of high pressure to such entities. With regard to the nature of the bonds involved in the antigen-antibody reaction and the susceptibility of these bonds to pressure), it can be assumed that the application of high pressure combined with a defined temperature operating window would lead, for example, to the separation of an antigen from its corresponding antibody.

Moreover, as mentioned earlier, the formation of affinity complexes is often accompanied by an increase in molecular volume. According to Le Cha™-telier's principle that governs the effects of pressure, application of pressure to these complexes favours the disruption of such entities. Briefly, if in a given physical environment a positive association volume is observed, then pressure-induced dissociation can occur and reciprocally.

Baro-immunodesorption was based on the BSA-anti-BSA monoclonal filter-linked antibodies system. A comparison between the stability of the immunoaffinity supports as a function of the elution procedure (i.e. high pressure or low pH) showed a greater stability of the affinity matrix after the application of pressures up to 200 MPa than after low pH treatments.

Up to 90% of bound antigen were recovered after four cycles of pressureinduced desorption. Such immunodesorption experiments were performed with two other model systems: the prostate specific antigen and epiglycanin and their related antibodies. In these cases, the antigen was adsorbed onto microtitration plates, and the amount of released antibodies was measured; here again, pressure acted favourably on the disruption of immune complexes.

The preceding methods dealt with monomeric antigens devoid of enzymatic activity. To develop a baro-immunodesorption method whose recovery yield was based on the measure of an enzymatic activity. The tetrameric L-galactosidase of Escherichia coli was chosen as the model antigen. This enzyme is easily quantifiable, widely used in molecular cloning experiments, and often served as a tag in recombinant protein expression vectors. Moreover, its purification is easy either by conventional or affinity-based methods.

First the stability of the antigen under pressure was determined. The binding ability of pressurised antibodies was assumed, either in a free state or bound to membranes. These two latter determinations have to be done as others reported discrepancies, i.e. free or bound antibodies may behave differently with regard to pressure application.

The feasibility of the procedure was shown first with immunoa ffinity membranes, and second with threedimensional matrixes (i.e. agarose and silica beads). In addition, the recombinant L-galactosidase used possesses a LytA tag, which allowed its fixation on a DEAE-cellulose resin; the high pressure process might then be compared to a tag-based desorption procedure.

Protein-ligand interactions were also studied by affinity electrophoresis under pressure, and reviewed by

Masson. This method is funded on the retardation of migration of molecules in an electrophoretic field due to their interaction with ligands immobilised within the separating matrix.

This can be carried out either on cylindrical or on slab gels. An illustration of the determination of the apparent dissociation constant of complexes formed between bovine serum albumin and an immobilised derivative of stearic acid is shown in Fig. 1.

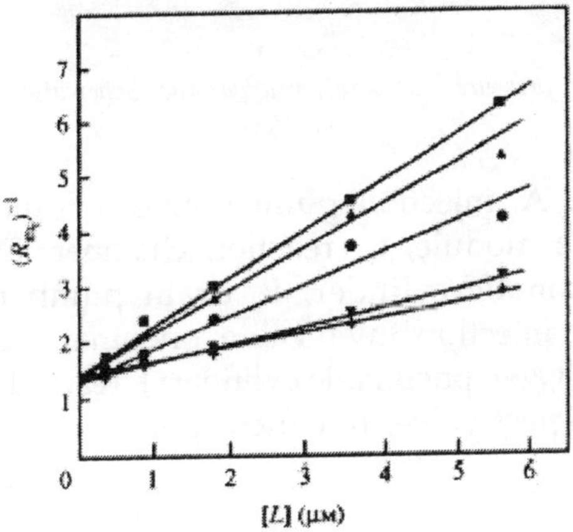


*Figure 1. Affinity electrophoresis.*

All the above presented methods are fundamentally discontinuous processes and need many manual operations. Two groups have proposed apparatuses allowing high pressure bioseparations in a continuous way.

The first one was tested on phage VDNA preparation, the separation of the RNA part of RNA/DNA hybrids, and the purification of an antibody following affinity capture by its corresponding antigen. This apparatus is presented in Fig. 2. The second one is a versatile plant whose functioning is based on percolation of exchangeable cartridges. The design allows to perform bioseparations, but also reactions of (bio)synthesis. This plant is shown in Fig. 3. The two devices are driven by LabView or LabWindows based computer software.

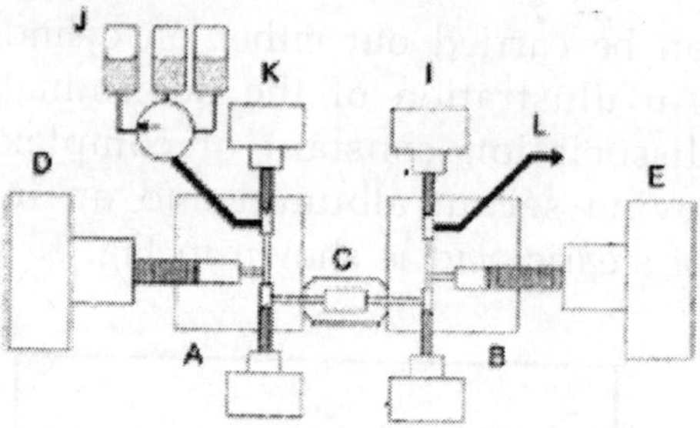


*Figure 2. High pressure bioseparation apparatus. Schematic of system and actuator.*

In figure 2, A, injection pump/valve module; B, drain pump/valve module; C, reaction chamber; D, injection pump pneumatic cylinder; E, drain pump pneumatic cylinder; F, injection inlet valve pneumatic cylinder; I, drain outlet valve pneumatic cylinder; J, reagent reservoirs; K, reagent select valve; L, e•uent port.

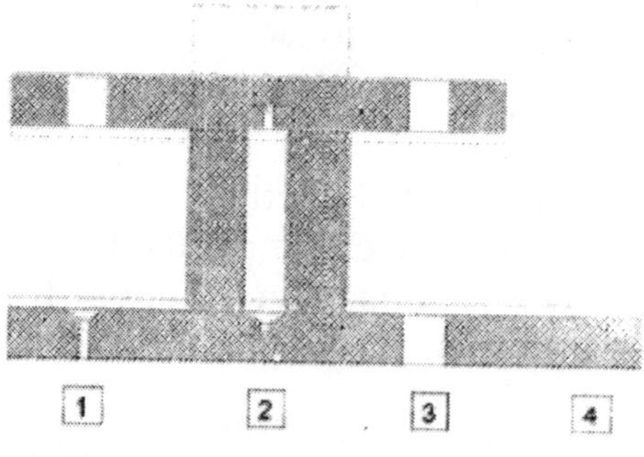


*Figure 3. Semi-continuous high pressure plant.*

In figure 3, this plant is computer driven and might be used either for (bio)synthesis and/or bioseparation processes. Its conception is based on a moving high pressure reactor containing removable cartridges. The reaction cylinder may occupy four positions permitting low (pos. 1), high percolation or reaction operations (pos. 2), change of cartridge (pos. 3), and maintenance (pos. 4).

In order to enhance the performance, in terms of specificity, safety, and yield, of a baro-immunodesorption process, one can act at different levels: the susceptibility of antigens to pressure, the generation of antibodies from non-animal origin, the development of new affinity matrixes, and the selection of (bioengineered) antibodies.

#### *Antigens and Antibodies*

The influence of pressure on antigens and antibodies were performed on anaphylaxis, antitetanic and antidiphtherial models as early as 1933. An approach combining fluorescence (to detect protein conformational changes), FTIR (to determine secondary structure alterations), and ELISA (to check antibody recognition ability) was suggested to screen antigens and antibodies as to their usability in an immunodesorption process.

Also, the adjunction of a metatype antibody (which recognises antibodies in a liganded state) to immune complexes stabilises this latter against the action of pressure. Besides the classical immunisation methods of animals with whole antigens, one of the 'oldest' ways to get specific antibodies looking like monoclonal ones is to raise antibodies against synthetic peptides after having determined potential antigenic sites on a given protein. The great advantage is that this protein has not to be previously isolated; moreover, it may be a putative one revealed by a DNA coding sequence. On the opposite the generated antibody could either be only sequence or structure specific and then could not recognise the original whole native antigen or recognise many others.

#### *Non-animal Origin*

Due to the possibility of contamination by pathogens or toxic agents some people are reluctant to use compounds of animal origin in the elaboration, production or purification of molecules, particularly in food or pharmaceutical science. To overcome this point, a recent approach consists in the selection of antibodies from phage display libraries. Strategies to design and optimise library selection have been reviewed and strategic flow charts have been elaborated. Antibodies can also be produced in plants.

Formerly developed to study plant metabolism or alter the expression of phytohormones (abscisic acid or gibberellin), the expression of antibodies (plantibodies) in plants presents many advantages: no use of high cost microbial fermentation techniques, no use of mammals, post-translational modifications achieved as in eukaryotic cells. This latter property has to be emphasised, as it has been shown that the effects of glycosylation, for example, are of importance with regard to antibody function.

Finally, if 'clean' antibodies have to be obtained, immunoglobulins can be selected particularly with staphylococcal protein A-funded methods or with peptides binding to the Fc portion and originating from a phage display peptide library. Nevertheless, it should be kept in mind that these antibodies would obey to a pressure sensitivity criterion with regard to their barodesorption usability.

#### *Affinity Matrixes*

An improvement would also be found with the development of fully synthetic antibodies: the molecular imprinted polymers (MIP). These matrixes present physical as well as chemical stability, they resist storage and give recovery yields as high as 99%. No living organisms are involved in their preparation, thus they may be considered as safe supports. Their properties render them of interest in view of the conception of new immuno-barodesorption processes.

***Matrixes for Baroprocesses***

The affinity constants, and also the kinetic parameters, have to be determined before using antibodies in a biotechnological process. Many biophysical methods for the determination of these constants, such as equilibrium dialysis, band shift, competition ELISA, fluorescence quench, isothermal titration calorimetry, or real-time interaction analysis, have been discussed. A simple technique based on the Mancini immunodiffusion technique was proposed by Kennedy and Barnes and applied to the selection of eluents for immunoadsorptive chromatography.

The application of affinity electrophoresis in this field is of interest, as already mentioned. Association constants in the range of $10^7$-$10^8$ $M^{-1}$ are suitable for barodesorption processes. All these determinations have to be made with respect to the reaction media; for example, another parameter involved in the sensibility of antigen-antibody complexes is ionic strength. A critical review on the combined effects of temperature, ionic strength and pressure on several antigen^antibody complexes is available.

The evolution of association volumes, salt bridges, aromatic stacking as a function of the physical parameters applied, as well as a thermodynamic analysis were reported. All the above mentioned gaits would lead to the achievement of highly specific, pressure-sensitive antibodies (in terms of biodesorption).

## Bioproduction, Biopurification and Bioseparation

Until now mainly purification methods involving antigen-antibody recognition were evoked. The application of high pressure can also be considered in integrative schemes regarding the synthesis, production and purification of bioactive molecules.

### *Design of bioactive Molecules*

Many reviews dealt with the stability of proteins as a function of pressure and particularly with that of enzymes. The affinity for substrate, reaction rates, and specificity of enzymes vary as a function of temperature, pressure, and solvent, thus it is advisable to use high pressure in the design or achievement of particular molecules. This can be achieved either with biocatalysts of mesophile or extremophile origin.

The use of hyperthermophilic enzymes, as well as the characterisation of deep-sea microorganisms have been reviewed. Studies have been done to explain the properties of thermozymes. On the opposite, there also exist cold adapted enzymes of potential biotechnological interest. Some other compounds such as natural catalytic monoclonal antibodies (abzymes), or molecularly imprinted materials leading to synthesis or catalysis (plastizymes) could have their activity( ies) modified or improved under high pressure.

It should be kept in mind that the application of high pressure alters the specificity of organic reactions, leading generally to asymmetric syntheses. This would be advantageous if this property was used in conjunction with the above mentioned biocatalysts to obtain a specific stereo-isomer of a given bioactive product.

Biopurification can result from a direct precipitation of a protein from a crude extract, particularly with the use of antibodies. A review on immunoextraction applications was recently published. It can be assumed that the use of high pressure to disrupt immune complexes thus obtained is of interest.

Besides the immunochromatographic techniques previously described, there are more 'classical' methods of affinity chromatography separation, the performance of which can be improved by the application of high pressure as an elution protocol, for example the elution of proteins separated on the basis of their affinity for DEAE groups, dextran supports, metal affinity chromatography, or simple hydrophobic interactions.

Some new pressure-resistant chromatographic supports seem to be promising. These are non-porous copolymerised particles of styrene, methyl methacrylate and glycidyl methacrylate or molecularly imprinted polymers (already successfully used in the separation of amino acids, peptides, proteins, receptors, and enantioselection of drugs).

The precise understanding of the nature of recognition between molecular imprinted polymers and their 'substrates', the rules governing the enantioselectivity of bioseparation, and the definition of the kinetic parameters involved in affinity interactions would be very useful to enhance the performance in the bioseparation field by adding the potential positive effect of pressure to their intrinsic properties.

Finally, the ultimate stage required for determining the three-dimensional structure of a molecule is its obtention in a crystallised state. Here again, pressure was shown to act on the nucleation and growth of protein crystals.

#### *Integrated Processes*

In the conception of a process leading to the production

of a bioactive molecule under high pressure, many things must be kept in mind (e.g. non-specific adsorption of proteins to matrixes ). In all cases, it is better to use immobilised reactants (i.e. enzymes are usually more active and stable when immobilised on a matrix); they are also more convenient to proceed. From the preceding, one can summarise as follows. Pressure may act at different steps in the development of integrated processes. Elaboration of the product

— During in vitro synthesis by assuming enantioselectivity of organic synthesis, a modulation or modification of the specificities of natural (meso-, baro-, thermo-, psychro-) originating enzymes or abzymes, a modulation or modification of the properties of synthetic catalysts (plastizymes etc.).

— During in vivo production in bacteria, yeast, animals and plants, by operating on the metabolism of natural or engineered produ^ts, or allowing an overproduction of recombinant product (barosensitive promoters etc.).

Purification of the product

— Facilitation of crystallisation steps.

— Non-polluting way to assume desorption from ((immuno)affinity) matrixes.

A scheme pre,senting a process where all the compounds (including satellite products) may be recycled is presented in Fig. 4.

In figure 4, (A) Scheme of an expression vector comporting an immunoafinity tag, a specific site of chemical or enzymatic cleavage and the product of interest. (B) Diagram showing a possible process allowing the recovery of the fused protein, the product, and the tag. All the steps might be done under high pressure.

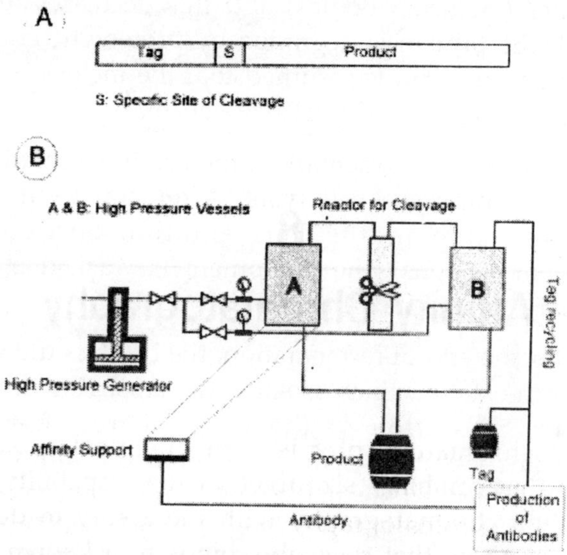


*Figure 4. Example of a biopurification process allowing the total recycling of all components.*

The antibodies are directed against the tag moiety of the fusion protein. In any case, it should be kept in mind that reusability of the various interveners is a fundamental criterion for their selection.

# 8

# Affinity Chromatography

Affinity chromatography is a biochemical separation method that combines size fractionation capability of gel permeation chromatography with the ability to design a stationary phase that reversibly binds to a known subset of molecules. These type of separations have greatly evolved over the past two decades to improve characteristics related to target specificity, dynamic adsorptive capacity, and chemical robustness of the affinity matrix. These separation matrices are used as screening devices for molecular interactions as well as for the purification of complex mixtures at the analytical, preparative, and large scale.

Affinity adsorptive phases can be synthesised by the attachment of structures, called ligands, to immiscible polymeric fluids, solvated gels, and porous solids. These ligands are selected for affinity properties for a soluble or colloidal target species. Advances in technology have been associated with the synthetic and biosynthetic design of affinity ligands and matrices. These advances have been made through a better understanding of phenomena associated with the transport of the target species to the ligand and the nature of molecular interactions between target and ligand. Specialised processing equipment for large and small scale applications have also improved affinity separations. This method is usually used to:

— Purify and concentrate a molecule from a mixture into a buffering solution.

— Reduce the amount of a molecule in a mixture.

Affinity Chromatography is based on the principle of biological recognition. Examples of biological recognition include antibody-antigen, enzyme-substrate, and receptor-agonist binding interactions. These interactions typically have high affinity ($K_d < 10^{-6}$ M), yet are reversible when conditions are changed. Due to the specificity of this recognition, it is often possible to obtain 100- or even 1000-fold increases in purity of the protein sample.

There are different steps in affinity chromatography. These are;

1. The first step in affinity chromatography is to prepare a stationary phase by immobilising one of the two recognised components on an insoluble hydrophobic polymer such as agarose.

2. A crude mixture containing the other recognised component is then applied to this stationary phase.

3. During a wash step, which often includes a high concentration of salt to overcome weaker recognition interactions, all proteins except those that bind tightly to the stationary phase are eluted.

4. The wash is followed by an elution step in which a soluble form of the immobilised ligand is used to displace the bound protein from the stationary phase. It may also be possible to release the bound protein by changing the pH of the buffers in the mobile phase.

There are thousands of examples of affinity chromatography in the literature, including the purification of trypsin and other serine proteases using immobilised bovine pancreatic trypsin inhibitor as the stationary phase.

Usually the starting point is an undefined heterogeneous group of molecules in solution, such as a cell lysate, growth medium or blood serum. The molecule of interest will have a well known and defined property which can be exploited during the affinity purification process. The process itself can be thought of as an entrapment, with the target molecule (ie the molecule targeted for purification) becoming trapped on a solid or stationary phase or medium. The non-target heterogeneous mixture will not become trapped.

The solid medium can then be removed from the mixture, washed and the target molecule released from the entrapment in a process known as elution. The process can be illustrated thus: Of course if the object is to remove an undesirable molecule from the initial mixture, then once the initial binding has been achieved, the material of interest is the unbound mixture, and the washing and elution steps can be omitted or used simply for analytical purposes.

### Uses of Affinity Chromatography

Affinity chromatography can be used in a number of applications, including nucleic acid purification, protein purification from cell free extracts and antibody purification from blood serum. Possibly the most common use of affinity chromatography is for the purification of recombinant proteins. Essentially proteins are tagged with a known affinity ligand in order to aid their purification. That is to say that the molecule has an artificial trait imposed upon it by genetic modification which allows it to be selected for affinity binding (if the molecule is a protein these are called fusion proteins).

Such tags include six His·tags (six Histidine residues added to one or both ends of a protein) and GST (glutathione-S-transferase) tags. Six His tags have an affinity for nickel ions which are covalently bound to NTA for the purposes of solid medium entrapment. For elution purposes a nickel chelating agent is used. GST has an affinity for glutathione (commercially available as glutathione sepharose). For elution purposes glutathione is used. Another use for the procedure is the affinity purification of antibodies from blood serum.

If serum is known to contain antibodies against a specific antigen (for example if the serum comes from an organism immunised against the antigen concerned) then it can be used for the affinity purification of the antibody. For example if an organism is immunised against a GST-fusion protein it will produce antibodies against the fusion-protein, and possibly antibodies against the GST tag as well. The protein can then be covalently coupled to a solid medium such as sepharose or agarose (these are commercially available). For thoroughness the GST protein and the GST-fusion protein can each be coupled separately.

The serum is initially allowed to bind to the GST affinity matrix, this will cause any anti-GST antibodies to bind (that is any antibodies against the GST part of the fusion protein will be removed). The serum is then separated from the solid medium and allowed to bind to the GST-fusion protein matrix. This allows the remaining antibodies to bind to the GST-fusion protein. Elution is achieved by the use of a low pH buffer such as glycine pH 2.8. The eluate is collected into tris buffer pH 8 as low pH will degrade the antibody. This is a nice example as affinity purification is used to purify the initial GST-fusion protein, to remove the undesirable anti-GST antibodies from the serum and to purify the target antibody.

### Column Chromatography

Binding to the solid phase may be achieved by column chromatography, whereby the solid medium is packed onto a chromatography column, the initial mixture run through the column to allow binding, a wash buffer run through the column and the elution buffer subsequently applied to the column and collected. These steps are usually done at ambient pressure (as oposed to HPLC or FPLC). Alternatively binding may be achieved using a batch treatment, by adding the initial mixture to the solid phase in a vessel, mixing, separating the solid phase (by centrifugation for example), removing the liquid phase, washing, re-centrifuging, adding the elution buffer, re-centrifuging and removing the eluate.

Sometimes a hybrid method is employed, the binding is done by the batch method, then the solid phase with the target molecule bound is packed onto a column and washing and elution are done on the column. The mechanism of affinity separations is most often described by a "lock and key" analogy. Retention depends on complementarity of a ligand molecule (the "key") and a specific binding site (the "lock") with high affinity for the ligand. Typically the molecule containing the binding site is a protein, although some nucleic acid interactions may also be useful. The ligand can be a small molecule, a carbohydrate, another protein, a prosthetic group, a specific amino acid sequence in a protein, or a specific nucleotide sequence.

Strategies imaginable for affinity separations are virtually unlimited, the only requirement being a strong and specific binding interaction that is in some way reversible. In some cases the ligand is covalently bound to the stationary media. The affinity adsorbent thus constructed can be used to specifically scavenge proteins or other macromolecules containing a binding site from a sample which may be a complex mixture containing many other non-binding molecules. This strategy is used to isolate proteins with binding sites for specific cofactors, enzyme inhibitors, cell-surface carbohydrates, and nucleic acid sequences, among other possibilities.

In other cases a protein with affinity for a specific ligand or sequence is covalently bound to the media and used to scavenge the ligand, or a molecule to which the ligand is covalently attached, from solution. Such is the case with protein A, originally isolated from Staphylococcus aureus but now most often produced by recombinant techniques, which has specific affinity for the Fc region of immunoglobulins.

Columns with protein A covalently linked to the stationary phase are routinely used for purification and analysis of monoclonal antibodies from ascites fluids or recombinant sources as well as mixed immunoglobulins from serum. Or the protein bound to the stationary phase may be an antibody. The column can then be used to pull proteins or other macromolecules containing a specific hapten (the binding moiety), or the small-molecule hapten itself, from the sample.

In operation, sample is applied to the affinity column with buffer and mobile phase conditions that favor strong binding of the ligand. After the sample has been applied, a brief wash with the sample buffer is performed to flush unbound solutes from the column. Then the mobile phase is changed to one that does not favor the ligand interaction and results in release of specifically bound substances from the column.

Mobile phases used for affinity separations are normally aqueous buffer solutions. The change that disfavors binding may be a change of pH. This is typically the case for protein A columns, where loading sample at neutral pH favors immunoglobulin binding and reducing pH to an acid value favors release. In other cases, the change that releases the binding may be a change in salt concentration, a temperature change, or introduction of mobile phase containing a competing ligand.

Historically affinity adsorbents have been constructed by covalent attachment of binding proteins or ligand to soft agarose or polymer gels. These support media have limitations. High flow rates and operating pressures cannot be employed because such conditions tend to compress the media and limit column lifetime. Also, historically, media with large particle sizes have been used, leading to low mass exchange rates. Flow, pressure, and mass exchange limitations all affect the speed at which affinity separations can be done and, as a result, the throughput available with a column containing a given amount of ligand or binding protein.

Column life and throughput are important considerations, especially for columns based on relatively expensive biochemicals such as protein A. Microparticulate wide-pore silica is an excellent medium for affinity separations, assuming the silica surface has been masked by appropriate treatment to eliminate non-specific interactions with ligands and binding proteins. Grace Davison's new ICEtech™ technology achieves this surface passivation and provides means for covalent attachment of a variety of ligands to silica.

The rigid pressure-resistant structure of silica overcomes the limitation of soft gel and polymer-based media by allowing columns to operate at higher flow rates. This contributes to the speed of separations and column longevity. The other factor that is important for high-speed affinity separations and high sample throughput is rapid exchange of materials in solution with those bound to the adsorbent.

High capacity in microporous adsorbents stems from their extremely high surface area per gram. The vast majority of the surface area, however, exists within micropores internal to the individual adsorbent particles. A solid adsorbent matrix which utilised only the area on the surface of its spherical particles would have very low capacity and limited utility for practical separations.

When sample is applied to a column under ligand-binding conditions, it is favorable for fast separations if molecules in solution can move rapidly to the interior surfaces of the adsorbent matrix. The rate at which this occurs depends on the diffusion rate of the molecules involved and the distance they must cover. The distance is inversely related to adsorbent particle size. Thus smaller adsorbent particles favor faster binding. Similarly during the wash period after sample addition, nonadsorbed molecules must diffuse out of the pores into flowing mobile phase between the particles in order to be washed from the column.

And during elution previously bound molecules, once released, must diffuse out of the pores into the flow stream in order to be cleared from the column and collected before the column can be re-equilibrated and the next separation cycle begun. The uniform small particle diameters of silica-based adsorbents complement their high flow and pressure stability, thus favoring rapid washing, elution, and fast cycle times for high sample throughput in affinity chromatography.

#### *Affinity Precipitation*

Main (Recombinant) proteins and increasingly also therapeutic DNA are among the most valuable products of the biopharmaceutical industry. The biggest challenge in the development of the related production processes is the development of an efficient strategy for product isolation (the downstream process). Such biomacromolecules are typically produced in a complex environment and initially 'contaminated' by many physicochemically very similar molecules stemming from the producing organisms or the culture medium. At the same time, the highest standards in regard to purity and safety have to be fulfilled in a product intended for the use in humans.

Speed, selectivity and preservation of the precious biological activity are therefore crucial in the development of the downstream process. The first step in downstream process development is typically the capture of the product-containing fraction. In the case of a technical enzyme, where up to 50 % of the expressed proteins may correspond to the desired product; this will often be an unspecific protein precipitation by ammonium sulphate. In the case of most pharmaceutical proteins or therapeutic plasmid DNA, where the product concentration may be below 1 % of the total protein / DNA fraction, capture has to be combined with specificity. Biological interactions such as antibody- antigen, receptor - ligand, are unsurpassed in their ability to recognise a specific biological molecule even in the presence of a host of similar ones.

Most production processes for recombinant proteins therefore contain an affinity step based on such a biospecific interaction for capture. Typically high purification factors are obtained in affinity chromatography. However, the use of chromatographic columns also poses problems during the early phases of the downstream process. The columns may clog, or may be difficult to regenerate and validate.

Affinity precipitation poses an alternative, which combines biological specificity with the simplicity of a precipitation. In affinity precipitation the biological ligand is linked to a stimuli-responsive material or molecule to create a stimuli-responsive specific capturing agent (stimuli-responsive bioconjugate). Such materials show pronounced property changes in response to a small change in an environmental parameter.

## Matrix Materials and Processing Geometries

Separations achieved by the selective capture of a target from a fluid phase by an affinity ligand confined to another adsorptive phase are most often done in aqueous systems. Therefore, the adsorptive phase for a hydrophilic target species is often a hydrophilic material. More hydrophobic targets require matrices which offer a more hydrophobic environment. In some cases, the adsorptive phase can be a separate fluid phase or soluble component having affinity ligands which will capture and concentrate the target species from complex aqueous mixtures.

Heterobifunctional ligands can bind the target and are also covalently coupled to a polymer which can be used to induce precipitation of a target-ligand complex. Alternatively, multiple but identical binding sites can be occupied on the target by homobifunctional ligands. The resulting crosslinked network between targets can be precipitated. Polymers used for affinity precipitation applications include chitosan, alginate, polyethyleneimine, and Eudragit S-100, which is a copolymer of methyl methacrylate and methacrylic acid. By manipulating such parameters as the pH or temperature, the polymer-ligand-target complex can be reversibly rendered soluble and insoluble. Lactate dehydrogenase and alcohol dehydrogenase have been precipitated using Cibracron Blue 3GA coupled with Eudragit S-100.

A significant advance in the field of bioseparations is the coupling of affinity ligands to aqueous two-phase systems of water-soluble polymers. Affinity partitioning is the selective extraction of target proteins from crude mixtures using affinity ligands immobilised onto either of two incompatible, water-soluble polymer phases. Mattiasson et al. used triazine dyes coupled to immiscible polymer fluids to purify proteins by affinity partitioning. Reactive triazine dyes have been widely exploited using this separation technique.

In comparison to classical chromatographic systems, affinity equilibrium is attained more rapidly in aqueous two-phase partitioning. In addition, high-affinity binding capacities and protein recoveries have been achieved. Alternatively to two-phase aqueous polymer partitioning, McCreath and Chase immobilised human immunoglobulin G onto perflourocarbon emulsions for the selective adsorption of Staphylococcus aureus cells containing membrane-bound Protein A. These liquid emulsion droplets were comprised of a perfluorocarbon oil that was crosslinked with poly(vinyl alcohol).

Most of the applications for affinity based separations are developed for chromatographic processing of aqueous systems. Thus, the ligand is immobilised on a porous bed of solids packed in a column through which liquid can be pumped. Most of the matrices are made of hydrogels composed of cellulose, agarose, and dextran. A chief advantage of hydrogel matrices is the combination of an easily derivatised, hydrophilic environment that provides molecular accessibility that can be extended to $10^6$ Daltons.

Other semi- and synthetic polymeric matrices, such as copolymers of vinyldimethyl azlactone and methylene-bis-acrylamide, can also provide similar intraparticle transport features. Coleman et al. immobilised protein A at high-density onto azlactone-functionalised polymeric matrices. These affinity matrices demonstrated high-throughput capacity combined with low operating pressures. As an alternative to the use of pellicular matrices or matrices comprised of reduced bead particle diameter, Rodriguez et al. and Liapis et al. described the mechanism of perfusion chromatography in which intraparticle mass transfer resistance is reduced by increasing the particle permeablity.

Velander et al. optimised the molecular accessibility and mechanical stability of uncrosslinked cellulose adsorbents by using largediameter beads ( > 0.3 mm) which operate at low pressures even at high flow rates. These hydrogel matrices had solids contents of only 3% or less and were shown to provide fast intraparticle transport, apparently through a mixed mode of both diffusion and convection. Larsson et al. demonstrated the use of superporous agarose matrices. The agarose beads contained a typical internal porous network in addition to larger pores. These larger pores constituted a significant portion of the bead, up to 1/3 to 1/10 of the bead particle diameter.

Mass transfer characteristics were improved since the larger pores allowed a considerable fraction of the bulk chromatographic flow to penetrate and flow through the individual bead particles. Affinity matrices were prepared using these superporous agarose beads containing immobilised $NAD^+$ analogue for the isolation of bovine lactate dehydrogenase and protein A for the isolation of rabbit immunoglobulin G. These affinity matrices operated under much higher processing flowrates compared to conventional homogeneous bead columns.

The protein A and $NAD^+$ analogue matrices had processing throughputs 5 and 3 times higher, respectively, in comparison to processing throughputs demonstrated by conventional affinity matrices. Thus, a significant improvement to affinity technology is the design of matrices having greater intraparticle accessibility and transport of the target species. Advances in transport phenomena needed for target contacting with the affinity adsorbent have resulted in matrix designs having large or small particulate as well as membrane geometries. Smaller media particles having mean diameters of about 0.1 mm or less yield a greater surface area to volume for improved target mixture contacting, but higher pressure drops when operated in a packed bed mode.

Thus, specialised affinity separations using high pressure liquid chromatography have been developed for small and large scale. However, these media require preclarification of the feed stream before chromatographic processing to prevent column fouling. Alternatively, small diameter particles having a higher density than water can be fluidised to achieve both filtration and affinity adsorption in a single step using expanded bed chromatography. Expanded bed adsorption chromatography has been developed to provide a single stage unit operation for the isolation of target proteins from crude mixtures such as milk, hybridoma cell culture fluid, and fermentation broth. For example, humanised $IgG_4$ antibody was isolated from myeloma cell culture fluid by expanded bed adsorption using recombinant protein A immobilised onto Pharmacia StreamlineTM media.

Planar membranes are often used for affinity separations in analytical assays but some cross-flow, stacked sheet geometries have been developed for large-scale applications. Large-scale affinity membranes used for large-scale applications can be also cast in tubular, hollow fibers. One example of planar membranes was made of a gel formed from two dispersed liquid phases. The chief advantages of affinity membrane systems are the low pressure operation and fast intraparticle transport due to short diffusion distances found in thin membranes. Thus, the kinetics of ligand-target interactions can become rate limiting to affinity separations using thin membranes. Etzel showed that limitations in membrane performance can arise due to variations in porosity and thickness which can result in diffuse breakthrough loading profiles. The stacking of planar membranes can sharpen affinity breakthrough profiles at higher loading flowrates.

**Affinity Ligand**

Affinity ligands have evolved from antibodies, enzymatic substrates, cofactors, nucleic acids, coenzymes, hormones, lectins, effectors, and inhibitors to a great diversity of small, low molecular weight peptides, polypeptides, and other organic structures. These newer classes of ligands can be made using biosynthetic and wholly synthetic methods. Common to all selection strategies is the need to begin with a diversity of structures from which to discover candidate affinity ligands. The chance discovery of a ligand with desirable properties has been enhanced through the use of phage display and synthetic combinatorial libraries which contain many orders of randomised structural permutations.

The structural permutations within a library rely on the polymeric assembly of bi- or multifunctional monomeric molecules into compounds which are typically greater than $10^3$ Daltons in molecular weight. In the case of biosynthetic libraries, the assembled structures are necessarily polypeptides and the subunits are amino acids. In the case of wholly synthetic structures, a variety of bi- and multifunctional organic subunits that provide high selectivity and yield reactions have been sequentially assembled using soluble or solid-phase linked chemistries. For example, various perturbations of difunctional molecules having reactive oxazoline chemistry have been used to create assembled combinatorial chemical libraries having $10^3$ or more structures.

In addition, some wholly synthetic libraries have used amino acids and novel branched chain peptide linkages as core structural platforms. In 1985, Smith demonstrated the use of phage technology to display candidate polypeptide ligands from a bacterial virus particle which also contains the DNA encoding the polypeptide sequence. The polypeptide is displayed from the virus particle surface in a manner that also enables affinity interactions with a targeted species. Thus, candidate ligands can be screened by the specific adsorption of phage to target species which have been immobilised to either a microtiter assay plate or to a chromatographic matrix.

Nonspecifically adsorbed material is then washed away and the specifically adsorbed phage particles are eluted. The eluted phage particles are cultured using microbiological platestreaking techniques. Candidate phages from the first affinity screening are passed through a second affinity screening and reculturing. Phages obtained from the twice affinity and culture selection processes are subjected to DNA sequencing. The DNA sequence of the displayed polypeptide is readily determined because of specific endonuclease restriction sites engineered into the phage genome.

The polypeptide ligand DNA sequence then is readily used to create ligands in mass quantity through recombinant fermentation technology. About $10^8$ polypeptides can be created within a single phage display library where about $10^1$ to $10^2$ candidate ligands typically result after a sequence of two affinity and culture screenings. Thus, the chief advantage of the biosynthetic, phage display libraries are the large number of library members and facile screening due to the coupled nature of the encoding DNA and displayed polypeptide ligand of the phage particle. The chief disadvantage is the inherent limitation to structures which are polypeptides.

However, as a further improvement of phage library technology, the use of random permutations about a core polypeptide ligand sequence can greatly enhance the affinity of initial ligand candidates. The chemical robustness of polypeptides is typically limited to moderate pH, and aqueous environments. Wholly synthetic combinatorial libraries typically use robotic assembly of structures. The robotic equipment used to generate orders of structural permutations essentially consist of a minaturised chemical factory. These multifunction automated work stations perform sequences of sample mixing, thermostatic reactions, volume reduction, and purification of the reaction masses.

The identity of individual reaction masses are catalogued. Identification of the newly created chemical structures are deduced from automated combinations of liquid chromatography and mass spectrometry. Libraries having $10^3$ to $10^4$ distinct members have been synthesised and characterised in this way. The screening of wholly synthetic libraries is laborious due to the need to putatively identify ligand-target complexes and to subsequently test the ligand structure in an immobilised state. In summary, a chief disadvantage of these libraries is the many order fewer library members that can be generated using the current robotic methods.

Another disadvantage is cumbersome information acquisition and management associated with both the characterisation of the ligand structure and target affinity. Unlike phage display, the primary molecular structure is not encoded and coupled to the wholly synthetic ligand as enabled by DNA of the phage particle. However, the diversity of chemically robust structures that can be produced from wholly synthetic molecules is an important advantage. In addition, rational design models can greatly decrease the number of permutations necessary from the initial discovery of ligands having lower affinity to the design of high affinity ligand candidates based upon the initially discovered, ligand structural motifs.

If combined with rapid and sophisticated assay screening methods, combinatorial chemistry methodologies provide an efficient process for the development of novel drug leads and affinity ligands. Future improvements in the efficiency of screening combinatorial ligands will likely be made using automated chemical sensor technology. Affinity peptide ligands can be made using solid-phase chemistry. Houghten et al. showed that the solid phase-linked chemistries have advantages of separation/wash cycling between reaction sequences as well asbeing able to provide information about the base structure of the candidate ligands.

Independent studies by Pingali et al. and Fassina et al. have demonstrated the use of affinity peptide ligands generated from solid-phase peptide synthesis. Pingali et al. produced an affinity peptide that bound human fibrinogen and a peptide of chromochrome c containing heme that captured human serum albumin (HSA). The anti-fibrinogen tetrapeptide was made using standard FMOC [N-fluorenylmethoxycarbonyl] chemistry. The affinity matrix made from this peptide was highly specific as purified fibrinogen was obtained directly from crude human plasma. Frontal analysis determined that the dynamic binding capacity of the anti-fibrinogen column was 10.2 mg fibrinogen/ml gel.

Similarly, by frontal analysis, an HSA-affinity peptide column was found to have a dynamic binding capacity of 19 mg HSA/ml gel. Fassina et al. used a core motif consisting of a small branched chain peptide to discover a protein A mimetic affinity ligand. This peptide recognised the Fc fragment of immunoglobulin G from rabbit, goat, sheep, and mouse. This affinity peptide provided a one-step isolation method for highly purified immunoglobulin G directly from crude sera. Frontal analysis determined that the dynamic binding capacity was 2 mg rabbit immunoglobulin/ml gel. This affinity peptide also exhibited stability towards sanitisation reagents such as 0.1 M sodium hydroxide and ethanol.

Both of the above mentioned studies show that peptide-based affinity columns derived from standard solid-phase peptide synthesis can provide a viable alternative to the use of immunoaffinity chromatography. Molecular modeling is used to deduce likely interactions between the target and affinity ligand, and is frequently used as a basis for further rational design of both biosynthetic and wholly synthetic combinatorial libraries. Once certain nominal affinity motifs have been identified from initial ligand screening, rational design has been used to enhance affinity by creating permutations which build upon structural motifs having lower affinity for the target species.

In addition, recent studies have been conducted which attempt to develop "peptidomimetic" structural compounds, such as aminimides, which have core structures similar to that of peptides, but with improved characteristics. Aminimides are much more chemically stable, have enhanced solubility characteristics, and are resistant to enzymatic degradation. Ringe et al. formulated a method for structure-based design involving aminimides in which complete binding surfaces of the targets are mapped.

Combinatorial chemistry is then used to identify and improve structural specifications in lead candidates targeted at the corresponding binding sites. Ringe et al. developed a peptidomimetic aminide inhibitor of porcine pancreatic elastase based upon crystal structures of an aminimide analogue. Affinity dyes are another class of wholly synthetic affinity ligands which are not derived from combinatorial libraries. Reactive synthetic textile dyes are resistant to chemical and biological degradation and can bind selectively and reversibly to a wide range of enzymes and proteins.

These ligands evolved from ligands which were first classified as enzyme cofactor mimetics. Many of these cofactor mimetics were members of reactive dyes used for staining textiles and paper. Since the mid-1950's, these reactive textile dyes have been used in the affinity purification of proteins. The use of commercial textile dyes as ligands in the large-scale processing of diagnostic, therapeutic, and genetically engineered proteins is documented throughout the literature. Since that time, many perturbations of original dye ligand structures, such as reactive Cibacron Blue, have been synthesised. These new structurefunction motifs are no longer considered a simple cofactor mimetic relationship, as was recognised between nicotinamide and Cibacron Blue.

Cibracron Blue F3G-A, the most thoroughly studied dye-ligand, binds with a variety of proteins such as adenine coenzyme dependent oxidreductases, phosphokinases, hydrolases, transferases, nucleases, polymerases, synthetases, lyases, decarboxylases, in addition to glycolytic enzymes, and plasma proteins. Lowe et al. has shown that the triazine dye structure to bind clefts of proteins which have no known cofactor binding domains, but have much more subtle, yet specific interactions with triazine dye derivatives.

Molecular modeling has also improved the rational affinity design of triazine derivatives through the addition of various chemical moieties to the core triazine structure. Lowe demonstrated the use of computer-aided molecular modeling and design in the development of structurally modified biomimetic dyes based on Cibracron Blue F3G-

A and Procion Blue MX-R. A dye ligand was engineered specifically for the capture of horse liver alcohol dehydrogenase. There are several advantages to the purification of pharmaceutical products using synthetic affinity ligands.

A major advantage is that they are not derived from biological sources. Therefore, they impart less of a product contamination risk. Hence, regulatory issues concerning the presence of unknown contamination and infectious agents in the final product are circumvented. Other major advantages of these ligands include: lower manufacturing cost; they can be readily immobilised under a wide variety of coupling chemistries to an extensive range of commercially available supports; they can be modified to enhance specificity or stability; they are more stable and less susceptible towards denaturation.

In addition, ligands derived from combinatorial libraries may contain a diversity of novel molecular structures such as organically derived or non-natural amino acid residues, such as L-amino acids (D-optical isomers). Well-characterised, affinity ligands may ultimately dictate decreased final product costs associated with less expensive process costs (i.e., more robust affinity-matrices, and less expensive ligands), and higher throughput due to a decrease in number of required chromatographic steps (i.e., decreased buffer usage). Affinity separation of pathogens may be better enabled by the increased accessibility of small ligands to subtle, yet conserved domains of viruses.

### *Installation of Affinity Ligands*

The covalent installation of the affinity ligand into a chromatographic or membrane-based matrix can profoundly effect the ligand-target binding efficiency. Matrix coupling chemistries usually covalently attach ligands through highly reactive groups such as a-amino groups. Common activation methods for polysaccharide matrices are cyanogen bromide, divinylsulfone, epoxy, organic sulphonyl chlorides, carbonyl diimidasole, and N-Hydroxy Succinimide. Polyacrylamide matrices are commonly activated by using glutaraldehyde or hydrazine. Isothiocyanate and $\gamma$-glycidoxypropylsilane activation methods are commonly used for silica-based matrices.

Ligands which contain amino acids can be coupled through the $\varepsilon$-amino group of lysine, carboxyl groups of aspartate and glutamate, and the phenolic group of tyrosine. Chemistries which randomly couple these residues also result in random orientation and spacing of the immobilised ligand. Some of the most thorough studies of ligand coupling chemistries have been done with proteins, especially antibodies. For example, Velander et al. demonstrated that antibodies used as affinity ligands exhibit best performance characteristics when the binding conformation of the antibody is protected during covalent coupling to the adsorbent phase. A conformationally related effect is the orientation of the immobilised ligand.

The use of oriented immobilisation methods for antibodies is also applicable to synthetic and biosynthetic affinity ligands that have asymmetric structure due to the presence of both target binding and nonbinding domains. The non-binding domain is best used for covalent coupling. Coupling of ligands to the support matrix through reactive moieties present within the binding domain can be detrimental from both orientation and nonnative conformational effects acting upon the ligand. However, masking or shielding of the binding domain prior to immobilisation can be employed to circumvent these effects.

Velander et al. used synthetic antigens comprised of water-soluble adducts of poly (2- methyloxazoline) polymers and a synthetic peptide epitope for the masking of monoclonal antibody during immobilisation. The mask was then removed from the immobilised antibody. A loss of as much as 50% of the theoretical binding capacity of an immobilised antibody was attributed to orientation effects based upon anti-Fc and anti-Fab probing of immunosorbents made using masked and unmasked antibodies. Activated matrices which preferentially couple through specific functional groups on ligands can aid in the site-directed, oriented immobilisation of affinity ligands. For example, Domen et al. developed several activated matrices in which each couple antibodies through different functional groups.

Iodoacetyl groups on SulfoLink gel are designed to couple through sulfyhdryl groups found predominately in the hinge region of antibodies. CarboLink activated gels couple through aldehyde groups. The aldehyde groups on antibodies can be formed by the oxidation of carbohydrate found primarily in the Fc region of antibodies. The site-directed method of coupling for a bivalent antibody through the oxidised carbohydrate groups using hydrazide chemistry resulted in the theoretical maximum of two antigen molecules for every antibody immobilised. However, Velander and Orthner found some antibodies immobilised onto matrices using hydrazide chemistry, which couple through the carbohydrate groups, produced no significant differences in immunosorbent antigen binding efficiency in comparison to immunosorbents prepared using random coupling chemistries such as cyanogen bromide. These antibodies were found to have carbohydrate in the binding domain.

The covalent attachment of the ligand through spacer arms becomes necessary when the nonbinding domain of the ligand does not offer sufficient molecular spacing between the molecular structure of the adsorbent and the binding domain to enable unencumbered target binding. The use of spacer arms are well documented throughout the literature. For example, Cuatrecasas demonstrated that extension of the ligand from the matrix through a hydrocarbon spacer arm can considerably increase ligand-target binding efficiency.

Cuatrecasas prepared specific Staphylococcal nuclease affinity matrices by immobilising the competitive inhibitor, pdTp-aminophenyl, using spacer arms of varying length onto agarose and polyacrylamide matrices. Affinity matrices prepared with the ligand attached directly to cyanogen-bromide activated Sepharose yielded a binding capacity of 2 mg nuclease/ml of gel. Whereas, affinity matrices prepared using a ligand attached to the support through a 3,3'-diaminodipropylamine spacer arm yielded a binding capacity of 10 mg nuclease/ml of gel. In addition to ligand proximity to the support surface, the proximal spatial positioning of adjacent ligands within the support influences binding efficiency as well.

A high local ligand density can also encumber target binding between proximally immobilised ligands. Velander et al. also demonstrated that antibodies immobilised with a low, local density gave as much as 50% of theoretical capacity based on a 2:1 target:antibody, molar stoichiometry. Thus, the majority of activity loss with associated with the installation of antibodies into matrices can be attributed to local density and orientation effects.

Intramatrix transport phenomena can also effect affinity sorbent performance, particularly for large target molecules. Thus, ligands are best installed into domains with rapid target accessibility. However, orientation, spacer arm, and local spatial ligand density effects must be evaluated at any location within the adsorptive matrix as part of the affinity sorbent optimisation process.

# 9

# Nanoscale Science and Engineering

Nanotechnology is an exciting and rapidly emerging technology allowing us to work, manipulate and create tools, materials and structures at the molecular level, often atom by atom into functional structures having nanometer dimensions.

Nature has been performing "nanotechnological feats" for millions of years. Through the arrangement of atoms and molecules, biological systems combine wet chemistry and electro-chemistry in a single living system. We are just beginning to catch glimpses of the nanoscale-methods used in nature to create self-replicating, self-monitoring, self-controlling, self-repairing, useful tools, materials and structures. "Nano" usually refers to a size scale between 100 and 0.1 nanometers (nm).

## Microfluidics

Fluids are used to do work for humans at the macro-scale. As an example, rivers and canal locks move boats and barges carrying people and supplies all over the world. On the micro-scale, movement of fluids through micro-sized channels permit controlled, precise manipulation of the fluid environment around delicate microscopic cells. On the microscopic scale, fluids in micro-channels are very viscous and move more like molasses than like water. When two or more microfluidic channels are converged into a single stream, the converging streams do not mix easily; this is a property of laminar flow.

Living cells can be arranged in single-file order using laminar flow of microfluidics, allowing new levels of precision science in cell biology. Other microfluidic applications are the potential for precision "point-of-care" diagnostics, drug screening, DNA manipulation and processing, monitoring of food, water supply, plant and animal health, and the environment.

Microfluidics is used today in animal science to significantly simplify traditional in vitro fertilization procedures used in animal breeding. Microfluidics also is an integral part of today's precision miniaturised "Lab-on-a-Chip" technology, allowing analysis and chemical manipulation of small samples. Microfluidics is at a larger scale than nanotechnology, but the merger of the dry world of nanotechnology with the "wet" world of biology will certainly involve the use of microfluidic techniques.

### BioMEMS

Methods of making micro-sized machines, or microelectromechanical systems (MEMS) are already established. Fully functional pumps, rotors, sensors and levers exist at the micro scale. Moving from the micro to the nano scale NEMS devices will present new engineering challenges. However, as with microfluidics, the integration of existing MEMS technology with biological systems will yield a new class of machines, "BioMEMS", that will perform functions at the nanoscale or on nanoparticles.

Biochips that are smaller than a postage stamp exist today. These biochips contain sealed channels and wells, electrodes for detection, connectors and fluidic input/

output ports. BioMEMS will provide the interface between the macro and nano-worlds, as well as the interface between the biological and electromechanical worlds of technology.

### Nucleic Acid Bioengineering

Many of the nanotechnology devices of today have been fabricated from larger pieces of silicon or other materials, through etching, much like sculpting away portions of the beginning material block to obtain the final structure. Nucleic Acid Bioengineering approaches manufacture from the opposite end - a "bottom-up" approach.

Nucleic acid engineering utilizes DNA molecules as building blocks, and forms specifically shaped particles that can be used to build larger units. Nanowires and nanomembranes from these basic building blocks are envisioned. Nucleic acid engineering is a platform technology that can find a myriad of applications for agriculture and food systems through the development of novel materials.

### Smart Treatment Delivery Systems

On the macro-scale, delivery systems such as the US Postal Service accepts items for delivery that are sealed in a package and have address instructions written on the outside of the package; both motorized and human carriers transport the package from one location to another. The development of "Smart Treatment Delivery Systems" on the nanoscale uses similar concepts applied at the molecular level.

For example, smart drug delivery systems in animals will most likely contain small, sealed packages of the drug to be delivered. The packages will not be opened until they reach the desired location. A molecular-coded "address label" on the outside of the package will allow the package to be delivered to the correct site in the body.

Nano and micro-scale mechanical systems may serve as the "carriers" in such a system. "Smart" delivery systems will also contain on-board chemical detection and decision-making capability for self-regulation that will deliver drug or nutrient treatments as needed. Remote activation and monitoring of intelligent delivery systems will assist agricultural growers of the future to minimize antibiotic and pesticide use.

## Nanobioprocessing

Bioprocessing is the use of natural biological processes to create a desired compound or material from a defined feedstock, such as compost material from plant and animal wastes. Nanobioprocessing will focuses on and utilize nanoscale technology to achieve the goal of bioprocessing with greater efficiency. The use of molecular probes or development of devices that allow rapid identification of microbes present in a feedstock are examples of research at the nanoscale that will increase the efficiency of bioprocessing. The product itself may be bulk or nanomaterial.

### *Bioanalytical Nanosensors*

Detection of very small amounts of a chemical contaminant, virus, or bacteria in agricultural and food systems are envisioned from the integration of chemical, physical and biological devices working together as an integrated sensor at the nanoscale. The bioanalytical nanosensors either use biology as a part of the sensor, or are used for biological samples.

### *Nanomaterials*

Nanomaterials can be materials that are either newly-created through nanotechnology, or that exist in nature, such as nanoparticles found in soil (clays, zeolites, imogolite, iron and manganese oxides) that provide the potential to manipulate structures or other particles at the nanoscale and to control and catalyze chemical reactions. Materials are composed of particles. The shape, structure and aggregation of particles at the nanoscale influence the properties of the material at the macro-level.

Nanoparticles have many applications in composite materials where they may provide transparency, or increased strength with decreased weight. "Smart fabrics" that can monitor the vital signs of the wearer are being investigated and is an example of some of the potential new uses and new processing methods envisioned for agricultural fiber products.

Soils are aggregates of nanoparticles, layer particles and organisms and water. Nanoparticles are also produced as agricultural by-products: airborne dust and aqueous runoff that cause air and water pollution. Controlling these nanoparticles is in the best interest of efficient, cost-effective and environmentally responsible agriculture. Viewing soil as a nanocomposite, and applying the paradigms and technologies of nanoscale science to it, will lead to more efficient and environmentally friendly agriculture.

### Bioselective Surfaces

Surfaces are the environment and location on which most chemical and biological interactions occur. A bioselective surface has either an enhanced or reduced ability to bind or hold specific organisms or molecules. Bioselective surfaces are important to the development of biosensors, detectors, catalysts, and the ability to separate or purify mixtures of biomolecules, as well as in the processing and packaging of food.

**Agriculture and Food Systems**

Nanotechnology has the potential to revolutionise agriculture and food systems in the United States as a new enabling technology. Agricultural and food systems security, disease treatment delivery systems, new tools for molecular and cellular biology, new materials for pathogen detection, protection of the environment, and education of the public and future workforce are examples of the important links of nanotechnology to the science and engineering of agriculture and food systems.

Some overarching examples of nanotechnology as an enabling technology are:

— Production, processing, and shipment of food products will be made more secure through the development and implementation of nanosensors for pathogen and contaminant detection.

— The development of nanodevices will allow historical environmental records and location tracking of individual shipments.

— Systems that provide the integration of "Smart System" sensing, localisation, reporting and remote control will increase efficiency and security.

— Agricultural and Food Systems security is of critical importance to homeland security. Nation's food supply must be carefully monitored and protected. Nanotechnology holds the potential of such a system becoming a reality.

Today in agriculture, if a plant or animal becomes infected with disease, it can be days, weeks, or months before the presence of disease is detected by wholeorganism symptoms. By that time, infection may be widespread and entire herds/fields will need to be destroyed. Nanotechnology is working at the same scale as a virus or disease infecting particle, and thus holds the potential for very early detection and eradication. Nanotechnology holds out the possibility that "Smart" treatment delivery systems could be activated long before macro symptoms appear.

For example, a smart treatment delivery system could be a miniature device implanted in an animal that samples saliva on a regular basis. Long before a fever develops, the integrated sensing, monitoring and controlling system would detect the presence of disease and notify the farmer and activate a targeted treatment delivery system. Smart treatment delivery systems are envisioned for biology and bioactive systems such as drugs, pesticides, nutrients, probiotics, nutraceuticals and implantable cell bioreactors.

The fundamental life processes of agriculture are explored through research in molecular and cellular biology. New tools for molecular and cellular biology are needed that are specifically designed for separation, identification and quantification of individual molecules. This is possible with nanotechnology and will permit broad advances in agricultural research areas such as reproductive science and technology, conversion of agricultural and food wastes to energy and other useful by-products through enzymatic nanobioprocessing and disease prevention and treatment in plants and animals.

New materials that have special characteristics at the nanoscale will be a tremendous breakthrough in agriculture and food systems for pathogen and contaminant detection. Materials that have self-assembly and self-healing properties will find a multitude of applications in agriculture. Packaging of food in self-healing containers would prevent food microbial contamination and facilitate food preservation, storage and distribution. Protection of the environment through the reduction and conversion of agricultural materials into valuable products is an exciting potential area of advancement made easier by nanotechnology.

The design and development of new nanocatalysts for the conversion of vegetable oils into biobased fuels and biodegradable industrial solvents is one approach already under scientific examination that would be greatly enhanced with the addition of nanotechnological abilities. Protection of the environment through management of local and environmental emissions is another exciting area of agriculture that could benefit from nanotechnology.

Before reaching the dinner table, the lettuce salad, baked potato, broccoli and warm wheat bread have survived a formidable number of challenges from the environment and the environment has imposed decisions on the grower of each crop regarding the optimal timing of planting, irrigating, fertilizing and harvesting. Agricultural crops must be protected against the invasions of wild animals, weeds, insect pests, fungal pathogens, and the whimsy of the weather.

Close daily scrutiny, or "scouting" of crops for potential problems is critical to the health of the crop and also reduces the amount of pesticides needed. The Integrated Pest Management (IPM) approach, widely adopted in US agriculture today, reduces pesticide use on plant crops by only applying pesticides when needed, as determined by scouting the crop for pests. However, scouting is a time-consuming task for the grower, and requires a significant degree of expertise to recognize and diagnose symptoms of plant problems from insects, fungal pathogens, or animal-damage.

Scouting for problems brought on by the environment also is important. The weather imposes challenges to plant growth through weather-induced water stress, heat stress, or through soil nutrient depletion. Timing and extent of irrigation or fertilization for various areas of the crop are determined by scouting. Many of these tasks could be simplified through nanotechnology. Computerized control of the environment over small parcels of land is known as "Controlled Environment Agriculture" (CEA).

CEA technology, as it exists today in the US, Europe and Japan, provides an excellent stage for the introduction of nanotechnology into plant production agriculture. With many of the monitoring and control systems already in place, nanotechnological devices for CEA that provide "scouting" capabilities could tremendously improve the grower's ability to determine the best time of harvest for the crop, the health of the crop, and questions of food security such as microbial or chemical contamination of the crop.

The National Nanotechnology Initiative's technical report describes the initiative and its implementation plan. The President's supplemental report briefly identifies several impacts and applications of nanotechnology for agriculture and biotechnology. Among the exciting potential applications are biosynthesis and bioprocessing of new chemical and pharmaceutical products, integration of biological building blocks with synthetic materials, imitation of biological systems, molecular-engineered biodegradable chemicals for nourishing plants and protecting against insects, genetic improvement for animals and plants, drug and gene delivery to animals, and nanoarray-based technologies for DNA testing.

### Nanosensors

Today, sensors provide an abundance of information about such parameters as temperature and weather data and data that provide information on air, land and sea transportation, chemical contaminants, deceleration for release of airbags in automobiles and countless other variables. Biological organisms also have the ability to sense the environment.

For example, humans sense the environment through sight, touch, taste, smell and sound. In living organisms, sensors operate over a range of scales from the macro (ear drum vibrations) to the micro (nerve cells) to the nanoscale (molecules binding to sensors in our noses). The exciting possibility of combining biology and nanoscale technology into sensors holds the potential of increased sensitivity and therefore a significantly reduced response-time to sense potential problems.

Imagine, for example, a bio-analytical nanosensor that could detect a single virus particle long before the virus multiplies and long before symptoms were evident in the plants or animals. Some examples of the potential applications for bioanalytical nanosensors are detection of pathogens, contaminants, environmental characteristics (light/dark, hot/cold, wet/dry), heavy metals, and particulates or allergens. Desirable characteristics of biosensors were identified by the panel as: small, portable, rapid response and processing (i.e., real-time), specific, quantitative, reliable, accurate, reproducible, robust and stable.

Several sub-groups discussed the specific objectives of a national research program in bioanalytical nanosensors. The combined recommendations are:

— to develop pathogen, nutrient, and contaminant detection systems and

— to obtain an improved understanding of nanoscale sensor mechanisms through fundamental research and modeling.

*Detection systems (Pathogen, nutrient deficiency or contaminant)*

— Sample Retrieval: Develop retrieval nanosystems for sampling of specific components (from air, plant and animal organisms, water, soil).

— Pathogen Detection: Develop methods of near real time pathogen detection and location reporting using a systems approach, integrating nanotechnology micro-electromechanical systems (MEMS), wireless communication, chip design, and molecular biology for applications in agricultural security (economic, agricultural terrorism, agricultural forensics) and food safety.

*Fundamental research and mechanistic modeling*

In order to design the best nanosensors, we must understand the fundamental ways things work when sensing at the nanoscale.

— New ways for nanosensors to capture and hold the pathogen or chemical of interest are needed. New immobilisation techniques might be based on chemical, biological, or electrical methods of capture. . New methods for sensors to "recognise" pathogens or chemicals must be found. The most promising areas for development of new recognition mechanisms are through fundamental studies in the areas of non-biomaterials, biomimetics, carbon nanotubes, molecular imprinted polymer (MIP), and recombinant/genetically engineered biosensors.

— New methods are needed for sensors to send a signal (also called "Transduction" of the signal) once it has recognised the pathogen or chemical of interest. Some promising possible methods for development of novel transduction mechanisms are mechanical, impedance, piezoelectric, optical, electrochemistry, transistor, DNA nanocircuits, and DNA resist/photolithography techniques.

— The above two steps (recognition and transduction of signal) must be integrated to work together. Promising areas for development of novel integration mechanisms between recognition element and the transducer may be found in utilization of the nanotechnology techniques of self-assembled monolayer (SAM) and directed/guided assembly.

### Identity Preservation and Tracking

Nanotechnology in agriculture brings possibilities that at one time were only possible in science fiction. George Orwell's futuristic science fiction novel "1984" imagined a surveillance system named "Big Brother" watching over all peoples and locations. Identity Preservation (IP) is a system that creates increased value by providing customers with information about the practices and activities used to produce a particular crop of food or other agricultural product.

Certifying inspectors can take advantage of IP as a more efficient way of recording, verifying, and certifying agricultural practices. Today, through IP it is possible to provide stakeholders and consumers with access to information, records and supplier protocols regarding such information as the farm of origin, environmental practices used in production, food safety and quality and information regarding animal welfare issues. Some food or processed agricultural products may be stored for years, with intermittent samplings for storage pathogens or environmental storage problems.

Each day, shipments of food and other agricultural products are moved all over the world. Currently, there are financial limitations in the numbers of inspectors that can be employed at critical control points for the safe production, shipment, and storage of food and other agricultural products. Quality assurance of agricultural products' safety and security could be significantly improved through IP at the nanoscale. Nanoscale IP holds the possibility of the continuous tracking and recording of the history which a particular agricultural product experiences. We envision nanoscale monitors linked to recording and tracking devices to improve identity preservation of food and agricultural products.

In setting the objectives of a national research program in this area, the panel focused on the "first steps" of IP for agriculture and food systems involving a measurable, realistically attainable goal, based on temperature changes due to changes in metabolism. Other environmental/physiological factors could be added to the system as the technology allows. Topics were identified as research priorities for a national research program in nanoscale identity preservation for agriculture and food systems.

Quantify metabolic process energetics at a macromolecular scale using biodegradable sensor devices. Develop a nanothermal device/data logger (monitoring temperature changes) for life history of agricultural commodities (storage, shipping, delivery to store and transfer to the consumer). Develop devices/data loggers for detection of pesticides and fertilizers for life history of agricultural commodities (storage, shipping, delivery to store).

### Smart Treatment Delivery Systems

Today, application of agricultural fertilizers, pesticides, antibiotics, probiotics, and nutrients is typically by spray or drench application to soil or plants, or through feed or injection systems to animals. Delivery of pesticides or medicines is either provided as "preventative" treatment, or is provided once the disease organism has multiplied and symptoms are evident in the plant or animal. Nanoscale devices are envisioned that would have the capability to detect and treat an infection, nutrient deficiency, or other health problem, long before symptoms were evident at the macro-scale.

This type of treatment could be targeted to the area affected. The development of "Smart Delivery Systems" for agriculture is defined as any combination of the following: time-controlled release, spatially-targeted release, self-regulated, remotely regulated, preprogrammed, or multifunctional so as to avoid biological barriers to successful targeting.

Smart delivery systems also could have the capacity to monitor the effects of the delivery of pharmaceuticals, nutraceuticals, nutrients, food supplements, bioactive compounds, probiotics, chemicals, insecticides, fungicides, or vaccinations to people, animals, plants, insects, soils and the environment.

The following areas are identified as research priorities for a national research program for agriculture and food systems smart treatment delivery systems research.

— Develop delivery systems for biological and bioactive systems (drugs, pesticides, nutrients, probiotics, nutraceuticals and implantable cell bioreactors).

— Develop integrated sensing, monitoring and controlling capabilities with onboard intelligence for self-regulation or remote activation for food production, storage and packaging applications.

— Develop targeted site delivery capability from implants in animals and plants that are activated only as needed.

— Design food nanostructure, oral delivery matrices, particulates, emulsions and nanodevices for enhanced food flavor and digestibility.

## Smart Systems Integration

Smart Systems Integration are Sensing, Localization, Reporting and Control. "Smart Systems Integration" is similar to designing and building the logic of a "nervous system" that will allow the individual parts to work together. Integration of the nanotechnologies into a working control system will require electronic communication between several technologies, including the sensing systems, reporting systems, localization systems and control systems. The logic to control the subsystems (control algorithms) must be developed and eventually translated into a computer language.

— Develop microfluidic devices that allow real-time temporal and spatial sensing and reporting for animal and plant production systems, food safety and security, biochemical/biomass processing or environmental monitoring applications.

— Integrate Nano-Electromechanical Systems (NEMS) with remote receive/transmit systems (embedded on the chip, satellite interaction, global positioning systems, remote powering, biopowering).

— Develop integrated sensing, monitoring and controlling capabilities with on board intelligence for self-regulation and remote activation for food production, storage and packaging applications;

— Design and develop automated integrated networks for monitoring and control of animal and plant production systems, food safety and security, biochemical/biomass processing or environmental monitoring applications.

## Molecular and Cellular Biology

Molecular and cellular biology hold the tools to understanding the most fundamental life processes in agriculture. Agricultural research in these areas has broad applications including reproductive science and technology, plant and animal breeding, conversion of wastes to energy and useful by-products, composting science and technology, plant physiology, veterinary medicine, plant pathology, disease prevention and treatment to name a few.

Three areas have been identified as research priorities in the development of nanodevices for molecular and cellular biology.

— Research and development of nanoseparation, identification and quantification devices is needed. Non-gel proteomic tools, membranes, sieves, and packings are needed to separate biomolecules in the range of < 100 nm in size. Tools for quantification also should be developed using fluorescent dyes attached to enzymes, nanoparticles, tags, markers, quantum dots, fiber optics or mass spectrometry.

— Nanobioreactor development is needed for the study of enzymatic processes, microbial kinetics, molecular ecology, mixed enzyme systems and rapid assessment of response to environmental factors.

— Research is needed in the development of nanodevices and materials for enhanced gene insertion processes, DNA delivery techniques for gene therapy, DNA vaccination, disease diagnosis and prevention for veterinary medicine and value-added plant and animal products.

### Nanoscale Materials

Nanoparticles are naturally occurring or human-made particles with nanoscale size or surface features that yield unique properties in materials at the macro scale. Nanoparticles currently have applications in imaging, printing, advanced materials, medicine, pharmaceuticals and drug delivery. The panel identified potential applications for and novel sources of nanoparticles in agriculture and food systems.

Novel materials developed through materials science and engineering are critical to the advancement of agriculture and food systems. Natural nanoparticles in soil, water, and air must be understood to the point that their characteristics and behavior can be controlled so that this natural resource may be more fully utilized. Agricultural practices also create and disperse nanoparticles. The environmental abundance of nanoparticles produced by agriculture must be understood and any negative effects mitigated.

— New nanomaterials development

  — Develop novel nanomaterials (DNA nanowires, DNA-microelectronic hybrids, bioseparation/ biofilms) by design using DNA as a building block. Building up from individual blocks rather than starting with a material and etching/carving away portions is referred to as a "bottom-up" approach to nanomaterial construction.

  — Investigate self-healing materials from self-assembled fatty-acids, surface modification (plasma technique) and coatings to modify properties of agricultural materials and packaging for the improvement of food packaging and prevention of food microbial contamination,

  — Develop better nanophase soil additives (fertilizers, pesticides, and soil conditioners).

— Materials science: bio-selective surfaces development

  — Develop surfaces with enhanced selectivity or that select against molecular interactions through biological, chemical, physical (including atomic, molecular, and nanoscales) for cells and biomolecules.

- — Develop smart surfaces that have controllable active spatial and temporal binding and release properties that are adaptable to agriculture and food systems (limited sample preparation, complex systems (dirty), more robust.
- — Develop novel bioselective surfaces (bind different molecules to DNA, DNA pattern at surface, porous metal with DNA, controlled pore size of DNA film, controlled molecular structure for filtration) using "bottom-up" approach.

— Fundamental nanomaterials science research

- — Conduct research to increase our understanding of the mechanics of nanomaterials; the transition from bulk to nanoscale behavior and chemical behavior in agricultural materials, changes in morphology, texture, chemistry.
- — Develop measurement methods and apply new measurement techniques to apply multiscale modeling to agricultural systems.
- — Characterize and model the physiochemical properties of natural and synthetic surfaces at the nanoscale to understand their basis for bioselectivity.
- — Characterize and model the process of self-assembly in biological systems such as triglycerides as templates for nano-self-assembly.

— Nanoparticles in soil and air

- — Understand the role of nanoparticles (inorganic and organic) in the transport and bioavailability of nutrients and pollutants.

- — Understand the transport and toxicity of nanoparticles in agricultural pollution (dust, feedlot runoff).
- — Understand soil as a complex nanocomposite, using the chemistry and physics of nanoscale and molecular phenomena to better model its properties.
- — Understand the role of nanoparticles in the global carbon cycle and the role of agriculture in global $CO_2$ levels.
- — Understand the role of nanoparticles in water retention and conditioning of soils.

— Biopolymers from plant waste

- — Identify new agriculturally-derived biopolymers for industrial and biomedical applications.
- — Explore more efficient methods of biopolymer modification; enhance processing for industrial products, harvesting, and fractionation.
- — Conduct research leading to a greater understanding of the structural and functional aspects of biopolymers.

— Nanocatalysts design for waste bioprocessing

- — Design and develop nanocatalysts for conversion of vegetable oils and other plant wastes (e.g., corn stover) to biodegradable industrial solvents and bio-based fuels.
- — Improve food quality through nanoscale processes that enhance the nutritional or nutraceutical composition of foods.

— Waste management and environment

Nanoscale processes or nanoscale phenomena that lead to the reduction and/or conversion of animal or plant waste into value-added products, and assist in the management of local and environmental emissions from agricultural or food production/processing systems.

Investment in nanotechnology materials research has the potential to reduce usage and costs of agricultural chemicals (fertilizers, pesticides), enhance environmental quality by detecting and mitigating environmental contaminants and provide longer shelf life and less contamination of food products.

### Educating Public

Nanotechnology is a part of nation's future and the future of agriculture and food systems. Thus, it is critical that the future workforce be trained in nanotechnology. The first step is informing the public at large about the advantages and challenges of nanotechnology for agriculture and food systems.

As the public awareness increases, so will interest in the study of nanotechnology by today's students increase. Transfer of technology to industry is another national priority identified by the panel and will further serve to educate the public and future workforce about the potential advances offered by nanotechnology for agriculture and food systems.

Nanotechnology is an exciting, futuristic, highly visible research frontier that can be utilized in education to excite students about agricultural science, food science, and agricultural and biological engineering careers. Showing the continuity between the fundamentals of nanotechnology and applied agriculture will draw attention to nanophenomena in agriculture, in nature, and will capture the imagination of young people considering careers in agriculture and food systems science and engineering. Agriculturalists of the future need a solid grounding in quantitative science and engineering. Research grants or fellowships for graduate study and research are a priority.

Nanotechnology and nanomaterials have the potential to play a significant role in risk reduction for issues of agriculture and food systems security. The public should be educated through television, Internet, and point-of-sale informative bulletins that explain the value-added, increased safety and food security due to application of nanotechnology. In addition to security of tracking systems and contaminant detectors, agricultural and food systems processes can be made more efficient, and reduce the extent of pesticide use, increasing food safety for the consumer.

The "dirty" nature of agricultural and food samples, it is recommended that separate, dedicated equipment be added to the existing centers for the analysis of agricultural samples. With the increase in industry working with academia and competing for government research funding, there is a need for restructuring of Intellectual Property issues.

# 10

# Biosafety

To assure safe processing, bioprocess engineers, operators, and managers must be intimately aware of the nature of the biohazard, the containment and regulatory issues, and how design and operation must satisfy the biosafety demands. The bioprocessing industry has an undeniable record of safe operation. Yet, equally undeniable is the continuing public concern regarding the safety of biotechnology, including bioprocessing.

Many hazards are associated with industrial bioprocessing: genetically modified organisms with real or perceived risks may have to be handled; highly pathogenic bacteria, viruses, and potentially contaminated substances such as blood may need to be processed; or the bioproduct may be so active that minute amounts may cause allergenic, toxic, or other activity associated reactions in personnel exposed to it.

## Risk Assessment

Most micro-organisms and animal cells used in industrial processes pose little or no risk to human health and the environment; nevertheless, some high-risk human, animal, and plant pathogens are used. By definition, a pathogen is any micro-organism or virus that can cause disease in any other living organism, including other microorganisms. Whenever a pathogen is used, it must be contained. In the long run, the trend is to move to safer processes by replacing harmful microbes with non-pathogenic recombinant producers.

The nature of the viable agent - microorganisms, viruses, animal and plant cells - has the greatest impact on the containment needs. Whenever possible, 'generally recognized as safe' or GRAS species should be used in commercial processes. Recombinant variants of GRAS organisms are preferable to non-GRAS microbes. When non-GRAS species must be used, known pathogens should be avoided, or variants not capable of producing disease should be preferred.

For example, pathologically incompetent Escherichia coli K-12 strains are used in producing recombinant proteins. Many microorganisms have a long history of safe use in food, and at least one previously unused species has been successfully commercialized as human food after exhaustive safety testing. Use of new species or those with unknown risks must be preceded by assessment of risk including pathogenicity testing.

Understanding of how viable and bioactive substances spread and invade the body is essential to assessing risk of processing. Microbial entry into the body occurs through inhalation of aerosols and particles; transfer to mouth via contaminated hands; damaged skin; and eyehand contact or splashes. In addition, bioactive substances may be absorbed through intact skin.

Aerosols and airborne particles are particularly troublesome sources of contamination. Aerosols spread easily and widely. Particles smaller than 5 μm dry instantly and remain suspended in the environment for long periods while circulating with air currents. Many operations generate aerosols, including centrifugation, homogenization, mixing, blending, aeration of liquids, leakage of liquids under pressure, and handling of solids. Laboratory procedures can contribute.

For example, aerosols are generated by bursting bubbles, breakage of liquid film as in pipeting, drops falling on surfaces, splashes, ultrasonic vibrations, sampling with syringe and needle, and during pouring and siphoning. Contaminated apparels, hands, equipment and process streams, and circulating air spread micro-organisms, as do insects, rodents, and other pests. Although relatively few infection episodes have been associated with industrial activity - the majority having occurred in research and diagnostic laboratories - only about 20 % of the laboratory-acquired infections have been ascribed to specific causes.

Unknown or unrecognized causes for most of the events suggest a continuing insufficiency of knowledge on the links between operational practices and infection. Thus, continuing vigilance is advised. Potential sources of contamination include direct accidental inoculation (needles, sharps, cuts or abrasions, animal bites), inhalation of aerosols, ingestion, and contact of contaminated material (hands, spills, contaminated surfaces) with membranes.

In addition to microbes, most physiologically active fermentation products - antibiotics, mycotoxins, enzymes, steroids, hormones, vaccines, deactivated microorganisms, antibodies, and other proteins - can be disruptive to health, and certain products are highly toxic. Aflatoxins are potent carcinogens. The fermentation conditions - temperature, pH, type of substrate, agitation, metabolic energy source, dissolved oxygen and carbon dioxide, the nature and concentration of micronutrients, metal ions, and other chemicals - influence the spectrum of biochemicals synthesized by an organism.

Under certain environmental conditions organisms such as Aspergillus flavus and Aspergillus oryzae are known to produce lethal toxins. Species of the genus Claviceps and some members of other genera produce toxic ergot alkaloids. Mycotoxin production is widespread among fungi. Several types of micro-organisms are known to cause allergenic reactions when inhaled in large amounts.

Allergenic reactions may be rapid, or the response may not occur until several hours after exposure, making connecting to the allergen difficult. Reactions may be extremely serious and, occasionally, fatal. Severe allergenic reactions to Bacillus subtilis proteases are well known. The hazard posed by nonviable bioactive material such as cytotoxic agents or endotoxins may not be eliminated by sterilization. In such cases, additional chemical decontamination of work areas, equipment and waste streams would be necessary using validated processes.

For example, solutions of sodium hydroxide (0.1 M) readily inactivate the botulinum toxin and are recommended for surface decontamination. Gram-negative bacteria produce thermostable endotoxins. Endotoxin-containing aerosols may be generated, for example, during cell disruption. Inhaled endotoxins are implicated in allergenic response; parenteral administration causes a pyrogenic reaction and other symptoms.

Adverse reactions to endotoxins have been observed with Enterobacter agglomerans, Flavobacterium sp., Methylophilus methylotrophus, Methylomonas methanolica, Pseudomonas aeruginosa, Serratia marcescens and E. coli. Up to 4% of the dry weight of E. coli K-12 has been estimated to be endotoxin. An action threshold value of 30 x 10-12 kg $m^{-3}$ for airborne endotoxin has been recommended. Increasingly, industrial processes use recombinant micro-organisms and animal cells. Biosafety considerations for solid-state fermentation processes have been discussed by Chisti, and issues relating to composting have been treated by Stentiford and Dodds. The commonly used bioprocessing schemes and individual unit operations have been detailed elsewhere.

### *Microorganisms*

Development of containment requirements for recombinant microbes must consider environmental and ecological consequences of inadvertent release. Important considerations include survivability and colonization potential of the organism in the environment, and the organism's ability to transfer any part of it's genome to indigenous populations. Survival and persistence studies are carried out in ecosystems such as activated sludge, mammalian gastro-intestinal tract, soil, and river water.

Gene transfer studies may be combined with those of persistence. Such work should be 'designed to show that the recombinant construct behaves similarly to the host in a representative ecosystem where the organism could be introduced inadvertently'. Assessments of potential biohazard should take into account characteristics of the unaltered parent, the unaltered plasmid vector, and the transposable elements. The U.S. Food and Drug Administration (FDA) has discussed these aspects in some detail.

As a general guide, a recombinant production strain should not have any known combination of pathogenicity, high colonization ability, and high genetic transfer competency. In addition to the genes of interest, selectable marker genes are introduced into the host during transformation. Safety of such markers and the proteins they encode remains a subject of debate.

Specifically, the effects of any antibiotic resistance markers should be considered: such markers may facilitate colonization of gastro-intestinal tracts of fermentation process workers receiving antibiotic therapy. Indeed, evidence is emerging that antibiotics in animal feeds alter the intestinal microflora in farm animals, and similar altered ecosystems become established in farm-workers that routinely contact the animals.

Recombinant E. coli K-12 strains and their plasmidless hosts are unable to establish in environments consistent with various deliberate release scenarios. Moreover, those strains are non conjugating and apparently incapable of transferring genes to other organisms.

### Animal Cells

Cell cultures may be contaminated with pathogenic viruses (e.g., HIV, hepatitis B) and mycoplasma (e.g., Mycoplasma pneumoniae). Again, immunosuppressed individuals are especially susceptible even to otherwise harmless viruses. Human cell lines are particularly high-risk and so are those derived from nonhuman primates; other mammalian cells are somewhat less risky, but may harbor agents capable of producing disease in humans.

Avian and fish cells, and those from invertebrates may be lower risk. Previously uncontaminated cells can become infected during processing through human contact, or use of contaminated sera. Other than being contaminated, established animal cell lines are potentially tumorigenic. Immunocompromised individuals are particularly susceptible, but a healthy immune system may be effectively circumvented if the transformed cell culture is compatible with an individual.

In one case, an accidentally inoculated (needle puncture, human tumor cells) laboratory worker developed a tumor. Clearly, as with any microbiological process, workers with temporarily damaged skin should not handle viable material. Because cell lines can harbor undetected viruses, and the lability of therapeutic proteins rules out the use of severe treatments that are capable of destroying viruses, a contamination-free product cannot be guaranteed; however, the risk of contamination can be reduced to extremely low levels using a multifaceted approach including use of exhaustively characterized cells and in-process controls.

Cells used in production originate in a manufacturer's working cell bank (MWCB) that is derived from a master cell bank (MCB). Only well-characterized cell lines are used in production. Characterization must assess identity of the cell line, it's microbial and viral contamination, genetic stability, and, for genetically modified cells, the genetic construct must be verified. Use of a well-characterized cell in production is insufficient assurance of a safe product; additional in-process controls are necessary.

Controls that need implementing include assessment of pre-harvest culture broth for relevant viruses and establishment of acceptance/rejection criteria for viral loads taking into account the validated capabilities of downstream virus removal/inactivation steps. Combinations of those approaches reduce risk to extremely low levels. In fact, so far not a single case of viral infection has been associated with the use of cell culture-derived biopharmaceuticals. Product purity and viral safety issues for cell culture-derived therapeutic proteins have been discussed further by others.

### Containment Requirements

Containment requirements are based on assessment of potential biohazard of an agent as reflected in it's risk classification. Conventional pathogens have been categorized into four risk groups with increasingly stringent containment needs.

A different biosafety level assignment is used for large-scale processing: GLSP, BLl-LS, BL2-LS, and BL3-LS

The BL1-LS (LS = large scale) corresponds to BL1 of the laboratory scale, and so forth. The GLSP is a lower level than the BL1 designation. The BL4 has no equivalent at the large scale, and agents requiring BL4 containment are not used in commercial production. BL4 organisms require high-level containment; they are pathogenic and hazardous to laboratory personnel; and they produce transmissible diseases for which no prophylaxis or treatment exists.

The biosafety level assignment considers whether an agent is pathogenic, poses a hazard to laboratory and plant personnel, is transmissible to the community, and availability of prophylaxis or treatment. Pathogenicity, or the ability to produce disease, depends on factors such as virulence, invasiveness, and infectivity. Of the four biosafety levels that are relevant to large-scale work, the GLSP level is suitable for non-pathogenic and non-toxigenic strains (including genetically modified vanants) that have an extended history of safe industrial use.

Ideally, all industrial processes should comply with those minimal requirements. GLSP is appropriate to organisms satisfying the following criteria:

— The host organism is non-pathogenic and free of adventitious agents. The host has an extended record of safe industrial use, or it has built-in incompetencies that limit it's survival in the environment, and it has no adverse environmental consequences.

— The genetically modified version is non-pathogenic and safe in an industrial setting, and without adverse environmental consequences. The containment levels shown are for relatively small-scale operations. For large-scale work, use of the next higher containment level is recommended (except BL3). Consult Richardson and Barkley for further guidance.

— The DNA vector is well-characterized and free from known harmful sequences. To the extent possible, the size of the insert is limited to that necessary for the intended function, and the insert does not increase the stability of the construct in the environment unless required by the intended function; the insert is poorly mobilizable, and it does not transfer resistance markers to micro-organisms not known to naturally acquire resistance if such acquisition would compromise use of a drug for controlling disease in man, veterinary medicine, or agriculture.

GLSP requirements further include:

— implementation of a health and safety program;

— suitably trained personnel and written operational procedures;

— facilities, equipment, protective clothing and practices appropriate to risk;

— regulatory compliance with regard to discharges to the environment;

— minimization of aerosol generation to prevent adverse risk to employee health; and

— a spill control plan within the emergency response plan.

Often, bioprocesses must comply also with the Good Manufacturing Practices (GMP) regulations; hence, in certain areas an otherwise GLSP process may actually meet much higher standards. For example, GLSP does not require any special containment, but to meet product protection requirements, use of enclosed equipment is advisable for processing of biopharmaceuticals.

Genetically modified Saccharomyces cerevisiae, E. coli K-12, B. subtilis, Aspergillus cryzae, and Chinese hamster ovary (CHO) cells, for example, can be used under GLSP classification. However, it should be noted that the bioactivity and the nature of the specific recombinant product being produced could strongly affect the acceptable containment level.

Thus, an otherwise GLSP species being used to make a toxic or unusually bioactive substance, may have to be contained not because of an intrinsic 'biohazard', but because of the nature of the product. GLSP-compliant processing aims to attain the lowest practicable exposure of workplace and the environment to any physical, chemical, or biological agent.

Processes requiring greater containment than GLSP should preferably be designed to one level higher than the minimum acceptable biosafety level. Typically, the cost of building to BL2-LS level is a minor increment over BL1-LS. Indeed, many U.S. companies have designed production facilities to BL2-LS containment requirements where the BL1-LS measures would have sufficed. Occasionally, the containment requirements for a given micro-organism differ slightly between jurisdictions. Again, the preferred practice is to err toward caution.

Depending on the risk, process support laboratories may have to comply with BL1-3 laboratory standards. Even a minimum containment GLSP process support laboratory should comply with good microbiological practices that are intended to protect both the operator and the product.

The EFB Working Party on Safety in Biotechnology continues to produce 'reports' that provide a useful perspective on biosafety issues. These reports have discussed plant pathogens, handling of micro-organisms of various risk classifications, assessing the impact on human health, hazard-based classification of micro-organisms, work with human and animal cell cultures, and general biosafety issues. The regulatory approaches to biosafety issues in the European Community and the United States have been compared.

In most countries, responsibility for biosafety issues is spread over several regulatory agencies. In the United States, biotechnology products and production facilities may come under the jurisdiction of the FDA, the U.S. Department of Agriculture (USDA), the Environmental Protection Agency (EPA), and the Occupational Safety and Health Administration (OSHA). The roles of the relevant agencies and the specific Acts under which they are empowered have been summarized elsewhere.

### Risk Management

Design and operation of a bioprocessing facility must assure safety of personnel within the facility and those in the surrounding community. Protection is achieved through a combination of engineered facilities, processes, and equipment; worker training and education; use of personnel protective equipment; operational practices; validation of machinery and methodologies; controlled access to facilities; biosafety committee or subcommittee; and medical and environmental surveillance.

Typically, enclosed process equipment or biological safety cabinets provide primary containment to the viable material. In the event of inadvertent release from process equipment, further secondary containment is provided by the building. Environmental monitoring seeks to ascertain the satisfactory functioning of primary and secondary containment. Here details the various aspects of risk management, including selection and use of biological safety cabinets which are invariably encountered at numerous stages of bioprocessing.

#### *Biological Safety Cabinets*

Biological safety cabinets are the primary means of

containment in process support laboratories and during early stages of culture development. Based on design and the protection afforded, biological safety cabinets are designated as Class I, II, and III.

Note that laminar flow "clean benches' are not biological safety cabinets and should not be used to handle potentially hazardous material. Class I cabinets (Fig.1) do not protect the work area against microbial or particulate contamination. The operator is protected so long as a minimum linear air velocity of 0.4 m $s^{-1}$ is maintained through the front opening. The cabinet is hard-ducted to the building exhaust system (Fig.1).

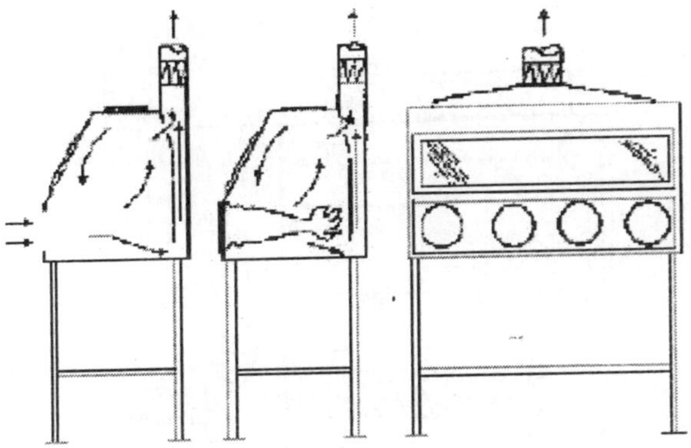

*Figure 1. Class 1, Biological Safety Cabinets*

Class II biosafety cabinets (Fig. 2) protect the operator, the product, and the environment. The work area is bathed in downward laminar flow of particle-free, recirculated air. In addition, air from the room is drawn in through the front opening to prevent leakage of aerosols and contaminated air. The linear air flow rate at the opening should be 0.4 m $s^{-1}$ or greater.

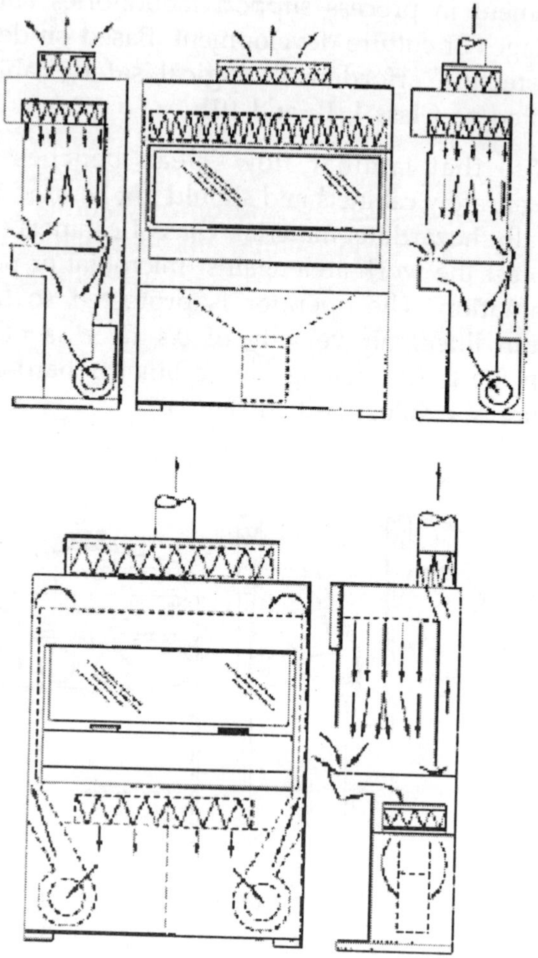

*Figure 2. Class II, Biological safety cabinets Types A (top) and B.*

HEPA filtered air may be exhausted into the laboratory as in Class II Type A cabinets (Fig. 2), or the discharge may be hard-piped to the building exhaust system as in Class II Types B 1-3. Type Bl cabinets recirculate part of the air over the work area, hence they may be used to process only minute amounts of volatiles. Type B2 cabinets are total exhaust devices that may be used also for some chemical containment, so long as the fumes are not susceptible to electrical ignition.

As a general rule, no class of biological safety cabinets is suited to handling volatile toxic substances, but nonvolatile toxic chemicals can be handled in all classes of cabinets Class II Type B3 cabinets are ducted Type A devices that like other Type B cabinets provide a minimum linear air velocity of 0.51 m $s^{-1}$ at the opening. All positive-pressure contaminated plenums within a Type B3 cabinet are surrounded by negative-pressure chambers to prevent leakage to the environment.

Class III biosafety cabinets (Fig. 3) are designed for handling BL4 biohazard agents. The cabinet is a fully sealed chamber with HEPA filtered air inlet and exhaust. The front end is provided with a sealed window and ports with heavy-duty, arm-length rubber gloves. Access to the chamber is through a side-mounted, disinfectant-filled dunk tank or through a sterilizable double-door pass-through such as an autoclave.

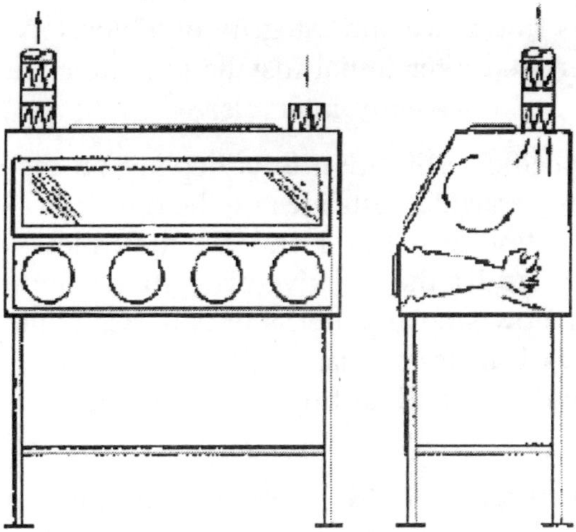

*Figure 3. Class III. Biological safety cabinet.*

The operator, the environment and the work area are protected. Air from Class III cabinets must be exhausted through two HEPA filters in series, or one HEPA filter and an incinerator. A dedicated, independent exhaust system exterior to the cabinet is used to maintain air flow. The enclosed work chamber is kept at a lower pressure than the laboratory.

Usually a 0.5 inch (1.3 cm) water gage pressure differential is maintained. Class III cabinets are usually installed only in maximum containment laboratories having other suitable safeguards. Biological safety cabinets rely on HEPA filters at air exhaust and/or intake to provide requisite protection to personnel, product and the environment.

Generally, HEPA filters are rated to remove particles down to 0.3 μm with an efficiency of 99.97 %, but more expensive higher efficiency (99.99 % or higher) filters are available. The 0.3 μm particles are least easily filtered compared with larger or smaller particles; hence that size is used for HEPA filter performance specifications. Filters are susceptible to shock-induced mechanical damage; therefore, performance and integrity of a biosafety cabinet must be certified after initial installation, after relocation, after repair, and at yearly intervals.

The specific certification tests depend on the type of cabinet. Some essential tests include the downflow velocity and volume testing (Class I and II); the inflow velocity test (Class I and II); the negative-pressure testing (Class II and III); air flow smoke patterns tests (Class I and II); the HEPA filters leak tests (Class I-III); the cabinet leak test (Class II and III); and testing of alarms and interlocks (Class III).

HEPA filters should be decontaminated prior to replacement. Decontamination is usually done with formaldehyde or hydrogen peroxide vapor, and decontamination provisions should be provided during installation. Class II cabinets usually have sensors for monitoring the pressure drop across the HEPA filter, and a low exhaust flow alarm is provided. Proper technique is essential for safe use of biological safety cabinets. The containment air curtain at the opening of Class II cabinets is easily disrupted by rapid sweeping arm movements into and out of the cabinet.

Other activities such as rapid movement of personnel around the cabinet and opening/closing of room doors also disrupt the air barrier. Arm movements into/out of the cabinet should be slow and perpendicular to the front face of the cabinet. The number of arm entries should be minimized by preparing a checklist of the required materials and placing them in the cabinet. The seating height should be adjusted so that the face of the operator is above the front opening.

Manipulations should be delayed for about 1 minute after arms/hands are placed inside the cabinet. Arms, hands or other objects should not rest across the front grill or the room air may flow into the sterile work area. All manipulations inside the cabinet should be at least 10 cm (4 inches) from the front grill. Any aerosol-generating equipment should be operated in the rear of the cabinet.

Used pipets should be discarded into a horizontal, disinfectant-containing discard tray kept within the cabinet. Potentially contaminated material should not be brought out until after surface decontamination with a suitable disinfectant. The cabinet air blower should be switched on at least 3-5 minutes before commencing work. The work surfaces and the interior walls should be disinfected before and after use by wiping with 70 % ethanol or other suitable disinfectant.

If chlorine bleach is used, a second wiping with sterile water is needed to remove residual chlorine, which is corrosive to stainless steel. Also, any material and containers placed in the cabinet should be wiped with 70 % ethanol to reduce risk of contamination.

#### *Spill Management*

Primary containment is provided by fully closed process equipment such as fermenters, centrifuges, heat exchangers, and pumps. However, even the best designed primary containment can fail; therefore, emergency response procedures must be in-place. Accidental release from process equipment poses a potential risk to employees and the local environment. The extent of risk depends on the pathogenicity, virulence, invasiveness, and infectivity of the agent, and the volume being handled.

Release or spill may occur within the confines of a biological safety cabinet, in unconfined areas within a facility, or more broadly due to catastrophic failure of a process equipment. For spills within a biosafety cabinet, immediate decontamination should be effected by a trained, suitably equipped (latex gloves, laboratory coat, safety glasses) technician while the cabinet air circulation system continues to operate.

Decontamination requires flooding the work tray with a disinfectant while minimizing aerosol generation. Thorough contact of the spill and the disinfectant is necessary for a preferred minimum of 30 minutes. The spill is then absorbed into disposable cloth or paper towels and discarded into autoclavable bags. The work surface, the cabinet walls, and any equipment inside is wiped with a disinfectant-soaked cloth. If the spill extends to the exhaust grills, the catch basin should be flooded (30 minutes) with disinfectant which is then drained into an autoclavable bag.

A disinfectant-soaked cloth is used to wipe the grill and the catch basin. The outside of the autoclavable container and bag should be wiped with a disinfectant-soaked cloth. Upon completion of cleanup, all solid material (including gloves, wiping cloth, lab coat, and any contaminated garments) that came into contact with the viable agent should be placed into an autoclavable bag. This material should be autoclaved at 121°C for a minimum of 1 h or other suitable period that has been previously established and validated for a particular load size and distribution.

Once contaminated gloves and clothing have been removed, germicidal soap should be used to wash arms, hands, and face. Disinfectants suitable for most purposes are chlorine bleach (500 ppm available chlorine), iodine solution (25-1600 ppm available iodine), formaldehyde (0.2 8.0 %), and 2 % glutaraldehyde. Only chlorine bleach is satisfactory for treating liquids; other agents noted are suited for wiping surfaces, glassware, etc.

Disinfectants such as quaternary ammonium compounds, ethyl alcohol, and isopropyl alcohol are not broadly effective. Chlorine bleach should not be used on stainless steel process equipment, and care should be taken to ascertain that bleach is compatible with the fluid being disinfected. The disinfection capability of chlorine bleaches declines with increasing pH. Readily accessible spill carts should be provided to deal with small unconfined spills.

A spill cart should have supplies of chlorine- and iodine-based disinfectants (e.g., 5 % Wescodyne, and 5 % Clorox), spill control supplies (autoclavable squeegee, autoclavable dust pan, autoclavable forceps, autoclavable biohazard bags, bucket, disposable wipes, and spill pillows), as well as protective clothing and equipment (disposable lab coats and jump suits, disposable latex gloves, disposable safety glasses, autoclavable boots, and half-face or full-face respirator with HEPA filter cartridges).

The following treatment procedure, adapted from Van Houten is recommended: Warn others of spill and leave the area holding your breath to avoid inhaling potentially hazardous aerosols; remove contaminated clothing, folding contaminated areas inwards and discard into an autoclavable bag; wash potentially contaminated body areas as well as face, arms, and hands with germicidal soap; shower if necessary; wear protective clothing (disposable lab coat, latex gloves, safety glasses, autoclavable boots).

If necessary, HEPA filter-equipped half or full-face mask; enter the area with the spill cart; use spill pillows to isolate floor drains if present and connected directly to sewer (i.e., not connected to a biokill treatment facility); encircle the spill with disinfectant, ensuring adequate spill-disinfectant contact while minimizing aerosolization; allow a 30-minute contact time; pick up broken glass and other sharp objects with the forceps, dust pan, squeegee, and place them in a leak-proof autoclavable container; use disposable wipes or spill pillows to mop up the liquid and discard it into an autoclavable bag; wipe the outside of the autoclavable bags with a disinfectant-soaked cloth.

Use an uncontaminated, disinfectant-soaked cloth to wipe the area of the spill. Upon completing clean-up, all solid material that came in contact with the viable agent should be placed into autoclavable bags. Bags, containers, and contaminated clothing should be decontaminated in an autoclave at 121°C for 1 hour or other prevalidated period. Arms, hands, and face should be washed with germicidal soap. Shower if necessary.

The spill cart should also be disinfected. Exposed personnel may have to undergo prophylactic or other treatments and medical surveillance in accordance with preestablished policies. Larger, unconfined spills are usually handled by especially trained spill response personnel using procedures similar to the ones noted for smaller spills.

### *Buildings and Facilities*

The design of a facility determines it's ability to provide secondary containment. Facilities processing especially hazardous material should preferably be located away from heavily builtup areas. In extreme cases, dispersal patterns for any inadvertently released material should be considered for all the meteorological scenarios relevant to the location. Access to the facility should be restricted to authorized personnel, with certain areas being 'off limit' to all but the relevant personnel.

Design concepts for biohazard containment

— Controlled access

— Work areas at negative pressure relative to surroundings

— HEPA filtered air exhaust

— Additional containment of aerosol-generating activities

— Personnel training

— Personnel protective equipment

— Decontamination of bioactive process wastes

— Medical surveillance of 'at-risk' personnel

— Environmental monitoring

Access control may require security fencing, electronic card controlled entry, vandal-proof exterior windows, etc. In addition to containment, the design of the facility needs to assure protection of the product. Products need to be protected against contamination, particularly microbial contamination to which they are highly susceptible.

Protection must be provided throughout manufacturing and storage. All facilities need an effective insect and rodent control program. Certain bioprocess facilities may require holding or handling of infected or potentially infected or suspect animals.

*Layout*

The layout of the facility affects efficiency of operations, the potential for containment, and prevention of cross-contamination. Attention to layout is required by the GMPs. Because the flow of personnel, equipment, materials and air in the facility must be controlled, the building and the process must be closely integrated by design. 'Contained', 'clean', and 'dirty' areas should be identified on the process flow sheets and the building layout drawings.

Movement of personnel, equipment, process streams, and air across containment boundary must assure integrity of containment through a combination of engineered systems and operational protocols. In general, flows must be unidirectional, from clean to dirty areas, and not vice versa. The dirty and clean paths should not cross, and there should be no back-tracking.

The biohazard containment areas and the aseptic product filling areas should be located in different wings of the facility, with no sharing of common hallways or direct access. Adequate space must be provided for various uses. There should be no overcrowding in work areas, especially the contained areas.

*Air handling*

Quality of the ventilation air and it's flow in a facility are crucial to containment, biosafety, as well as protection of the product. Air in a facility is handled by the heating, ventilation and air conditioning (HVAC) system.

Designing the HVAC system requires classification of process areas as 'clean', 'dirty', or 'contained'. Air-locks are used to isolate the contained areas and those with critical cleanliness requirements, from other zones.

Effective containment depends critically on management of air flow and pressure differentials. Area pressure differentials help to prevent airborne contaminants from intruding into contamination-free parts of a facility. Pressure differentials of 1.3 mm (0.05 inches) of water are typically used between adjacent areas. Areas containing infectious agents (e.g., viral vaccine) must be maintained under negative pressure relative to the surroundings.

Otherwise, the air flow is generally from 'clean' to 'dirty' areas. Access to a contained process area should be through an air-lock that is maintained at a lower pressure than the contained area and the outer access corridor. Access should be restricted (e.g., cardcontrolled entry). The HVAC system design should assure that unwanted air pressure differentials do not develop in the event of mechanical failure.

Visual and audible indication of ventilation failure should be provided. When feasible, once-through ventilation is preferred to prevent spread of contamination or the likelihood of cross-contamination. Air from the facility is exhausted usually on the roof, away from any air intake. All air from contained areas in a BL3 -LS facility should be exhausted through HEPA filters, and the area should be under negative pressure with respect to the surroundings.

Primary containment must be achieved by using closed systems or appropriate biosafety cabinets. The requirements notwithstanding, HEPA filtered air supply is recommended for minimizing the potential for contaminating the product. Pipework should be minimized in contained areas, and wall penetrations and light fixtures should be sealed.

Aerosol build-up in work areas can be reduced by HEPA filtered ventilation with a sufficient number of air changes per hour - 20-30 are not unusual in BL2-LS processing areas. The mandated minimum number of air changes in various areas may have to be exceeded to account for factors such as humidity and heat rejection in the area, it's typical function, personnel capacity, production of vapors and fumes, and generation of aerosols.

The relative humidity in most processing areas is controlled at 40-50 ± 5 %. Relative humidities exceeding 50 % promote corrosion, whereas values lower than 40 % lead to problems with static electricity. Ventilation is discussed further by First, del Valle, and Lee. Air quality is of particular concern in the processing environment of the sterile final dosage forms of pharmaceuticals. Areas where the sterile product, containers, and closures are exposed to the environment are designated as 'critical' areas.

For a long time, the air quality in critical areas was required to be at least Class 100, that is, no more than 100 particles of ≥ 0.5 μm per cubic foot of air. Higher standards are now in demand. In critical areas, the number of colony forming units (CFU) should not exceed one per cubic foot (0.03 $m^3$) of space, and a 0.05-inch (1.3 mm) water gage positive pressure differential must be maintained relative to adjacent areas.

Class 100 areas are typically contained within Class 100,000 areas in which the particle count of ≥ 0.5 μm particles does not exceed 105 per cubic foot (0.03 $m^3$). In addition, the CFUs do not exceed 25 per 10 cubic feet (0.3 $m^3$) of air. Moreover, the Class 100,000 surrounding space must equal or exceed 20 air changes per hour.

A higher number, generally 60-75 air changes per hour, must be provided in critical aseptic fill rooms. The Class 100 area itself should be designed for at least 600 air changes per hour. HEPA filtered air is almost always supplied at the ceiling, and low wall-mounted returns are preferred. Aseptic filling areas need to be maintained at positive pressure. Additional issues relating to HVAC system design for protection of the product have been discussed by del Valle and Dobie.

*Construction, finishes and practices*

How easily and well a facility may be cleaned, sanitized or decontaminated depends on it's construction, and finish. Processing areas should have non-shedding, smooth, impervious, splash-resistant finishes that are capable of being cleaned and disinfected.

Even a GLSP-level fermentation area should be capable of being hosed down. Contained areas should be capable of being sealed and decontaminated by disinfectant spray and by fumigation. Decontamination procedures should be established beforehand and validated. Formaldehyde vapor is a commonly used fumigant, but it is carcinogenic.

In addition, mixing of formaldehyde with chlorine-containing disinfectants can produce potent lung carcinogens. Paraformaldehyde gas may be used to surface-sterilize heat-sensitive equipment. Floors must slope to drains. Drains should have a slope of at least 2 cm per linear meter to assure complete drainage. Floor drains in areas that are susceptible to spills should not be connected directly to the municipal sewer; instead, the drains should be piped to the biokill system. Drains should be provided with 20 cm water trap seals to prevent back-escape of vapor, gases, and aerosols from the containment sump into the work area.

Sometimes, in addition to water traps, the contained drains are provided with check valves that prevent back-flow in case of a pressure build-up in the containment sump; but usually suitable venting of sumps and tanks is sufficient to preclude pressure build-up. All traps should be decontaminated after a spill, and on a regular basis by pouring several trap volumes of a disinfectant solution down the drain.

Catastrophic failure of pressure vessel fermenters is unlikely, but large leakages of fluid may occur from failed valves, ports, or gaskets. All material released up to the full volume of the largest fermenter should be contained within a diked area emptying into a sump, and treated through the biokill system. After the spill has been treated, the diked area should also be disinfected. The spill containment and treatment system should be capable of functioning on demand, and it should have sufficient capacity to handle the entire process volume.

A minimum containment capacity of twice the production capacity has been recommended which is quite reasonable considering that the process fluid as well as the subsequent wash effluent must be contained. Generally, concrete floors with troweled-on epoxy finish are preferred in processing areas, but in some low-traffic laboratory areas welded PVC flooring (not tiles) may be used. The epoxy finish (or vinyl) should extend at least 10 cm up the walls, and the edges should be coved.

Walls are generally concrete masonry units with epoxy paint. Ceilings are suspended dry wall, painted, or epoxy finished. In aseptic areas and those processing BLl-LS or higher agents, all internal corners (including wall-to-ceiling) should be preferably coved and all finishes should be flush (base flush with wall, door and window frames flush with walls). Sealed windows are the norm. In addition, any exterior windows should be break-resistant whenever high-level containment is required.

Light fixtures should be flush mounted, and, in high-level containment areas, they should be of a type that can be serviced from outside the contained area. The area furnishings should be of a sanitary design, resistant to water, process chemicals, and disinfectants. Usually, work surfaces are stainless steel or epoxy tops with baked epoxy painted casework. Placement of equipment and furnishings should not interfere with cleaning and disinfection.

Equipment may be placed on housekeeping pads with radiused edges, or raised on legs that comply with hygienic design standards. Floor penetrations should be minimized and penetrations should be fully sealed to prevent seepage. Supporting equipment from wall-mounted brackets, or from overhead supports is preferable. Hand-washing facilities should be provided near exits in the contained areas.

Sink faucets (taps) should be automatic, or elbow- or foot-operated. Suitable disinfectant soap should be provided. Although hand air dryers have been recommended, paper towels are preferable. Electric dryers recirculate particles and aerosols that are deposited on hands. A BL3-LS contained area should have separate facilities for gowning and washing at each entrance and showering facilities should be provided in close proximity.

Air-locks should be provided at all entrances and exits (including emergency exits) in a BL3 -LS contained area. Only a minimal number of essential personnel should be allowed into contained areas. Monitoring should be from outside the contained area, through sealed windows, intercoms, and closed-circuit television.

Containment features for pilot-scale fermentation facilities for producing cytotoxic agents and oncogenic viruses have been described. Some of the methods noted are no longer state-ofthe-art, but are effective nonetheless. For example, more elegant and reliable contained sampling methods are now available. Specific construction details of animal cell culture facilities have been noted by Donnelly and Lubiniecki.

The GMP regulations require provision of separate facilities for handling of spore-forming microorganisms. In addition, dedicated, segregated facilities are needed for processing penicillins (and other β-lactams) because cross-contamination with penicillins and penicillincontaining substances cannot be reasonably prevented in a multiproduct facility.

Separate air handling systems are necessary if a building processes penicillins as well as non-penicillin products. Similarly, in facilities producing several viral vaccines, Damm recommends complete isolation of areas dealing with different viruses. Containment and decontamination capabilities must function under normal conditions as well as during emergencies.

Provisions and practices must be in-place for evacuation during emergencies such as fire. Consequences of loss of power on containment should be evaluated during design, and emergency power should be provided to prevent loss of containment. In addition, some essential process equipment may require standby power and steam supply.

### *Process Equipment*

The design of process machinery determines it's primary containment capability. Moreover, specific equipment may have specific hazards associated with it's use. For example, highpressure equipment such as cell disruptors may generate sprays of contaminated fluid in the event that a gasket fails. Similarly, aerosols are produced during centrifugation and submerged aeration of culture broth in fermenters. Design and evaluation of process equipment require identifying areas where potential leakage of contained material could occur.

Points of possible leakage should be minimized, better contained, or, when feasible, eliminated. Containment capabilities should be assessed for all equipment, including fermenters, centrifuges, filters, solvent extraction units, cell disruption devices, spray and freeze dryers, sterilizers and autoclaves, pumps, valves, pipes, and heat exchangers, bottling and vialing machinery, downstream purification units, and waste treatment systems.

In addition, the HVAC system and the utilities should be assessed for potential of becoming contaminated. Potentially contaminated lubricating fluids, steam condensate, wash fluids, etc., should be treated through the contaminated waste system. Process equipment, control cabinets, electrical housings, switches, etc., should be splashresistant.

Alternatively, control cabinets and instrumentation may be located outside the contained area and serviced through sealed cables. Utility lines (e.g., steam, air, water, vacuum) that are connected directly to process equipment are at risk of becoming contaminated. Protective measures include maintaining a positive pressure in the delivery lines relative to contaminated equipment, use of microbial-grade filters and back-flow preventers.

To the extent feasible, equipment entering and leaving the contained area should be decontaminated by thermal sterilization in double-door autoclaves connecting the inside and the outside of the work area. Waste should be sterilized in a separate autoclave that is not used for process sterilizations. Autoclaves should have interlocked doors, and fail-safe devices that prevent opening the autoclave until a complete sterilization cycle has been implemented. Pre-vacuum type autoclaves are preferred for sterilization of solid waste and other general process uses. Sterilization cycles and load configurations should be pre-validated. Principles of thermal inactivation have been described elsewhere.

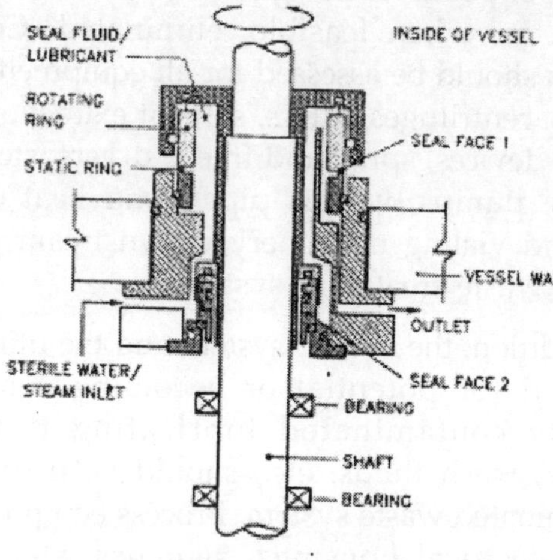


*Figure 4. Double mechanical seal with water lubrication.*

*Fermentation plant*

Fermenters usually contain large amounts of potentially hazardous viable material. Frequently, in addition to containing the viable agent used in production, entry of any other viable material into the fermenter must be prevented. Leakages from fermenters are not uncommon and considerations relating to containment and treatment of spills have already been discussed.

In addition, all through operation, most fermenters need to be aerated, and, the fermenter exhaust gases must also be sterilized. Usually, two 0.2 μm rated absolute hydrophobic filters in series are used to filter sterilize the exhaust air. Those filters are often heated to above dew point to prevent condensation, and, on larger or highly aerated vessels, filters are preceded by condensers. In addition, cyclonic separators installed before the filters may be used to protect the filter against contamination by foam and spray.

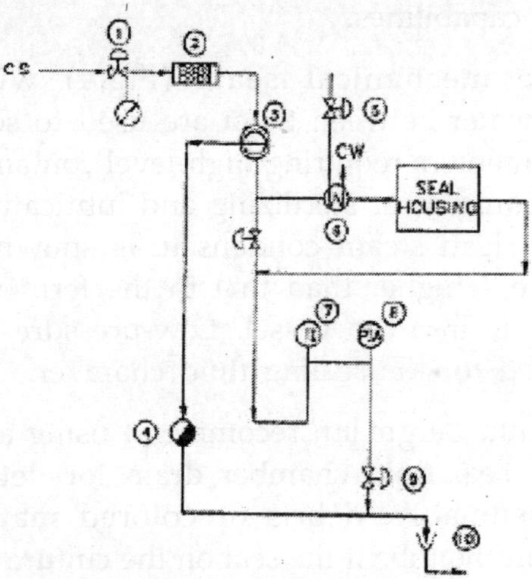


*Figure 5. Pipework for sterilization and sterile lubrication of mechanical seals.*

Mechanical foam breakers may also be used. In some cases, one exhaust gas filter may be followed by an incinerator. Integrity of the exhaust filters should be checked in-place using the forward flow diffusion method after the filter is sterilized, but before use.

Fermenters are protected against overpressurization with a rupture disc that should be piped to a HEPA-filter-vented containment/treatment system. Sometimes, the rupture disc is followed by a pressure relief valve so that the vessel returns to a contained state once the pressure is released.

This arrangement should have an attained pressure indicator for detecting disc rupture. In addition, use of overpressure sensors to shut off the sparger air supply in case of overpressurization has been recommended, but this practice is unusual: normally, the air supply regulator is set to a pressure significantly lower than the value required to open the rupture disc; thus, in case of a blocked exhaust filter, the vessel would attain a pressure equal to the air supply pressure that is still well within the vessel's capabilities.

Double mechanical seals (Fig.4) with sterile lubricating water between them are used to seal agitator shafts in fermenters requiring high-level containment. The pipework required for sterilizing and lubricating the seal with sterile clean steam condensate is shown in Fig. 5. (Fig. 4) is kept higher than that in the fermenter, hence any leakage is into the vessel. Low-pressure alarms are recommended for the sealing fluid chamber.

Miller and Bergmann recommend using a 'collection tube' in the seal fluid chamber drain for detecting seal failure. Presumably, debris or colored matter would accumulate in the tube if the seal on the culture side failed. A conductivity sensor in the water chamber should provide a better method of detecting leakage of culture fluid into the chamber.

Most fermentation media are relatively conductive, whereas uncontaminated sterile water produced by condensing clean steam is a poor conductor of electricity. Rotating seals can be eliminated altogether by using magnetically coupled agitators, but torque consideration limit the vessel size to about 800 L with animal cell culture bioreactors, and only to about 80 L with microbial fermenters. Large airlift bioreactors that do not require mechanical agitation are particularly suited to containment.

In addition, airlift reactors are more reliable than the mechanically stirred ones. Added protection against microleaks of viable agents can be provided by specifying double O-ring seals on all fermenter entry ports. In situ sterilizable probes that can be removed from a fermenter during cultivation, and decontaminated in-place before exposing the surroundings are available. The fermenter flanges may have double O-ring gaskets with a zone of live steam in between. Similarly, a live steam barrier can be maintained on the outside of valves that connect directly to the fermenter (Fig. 6).

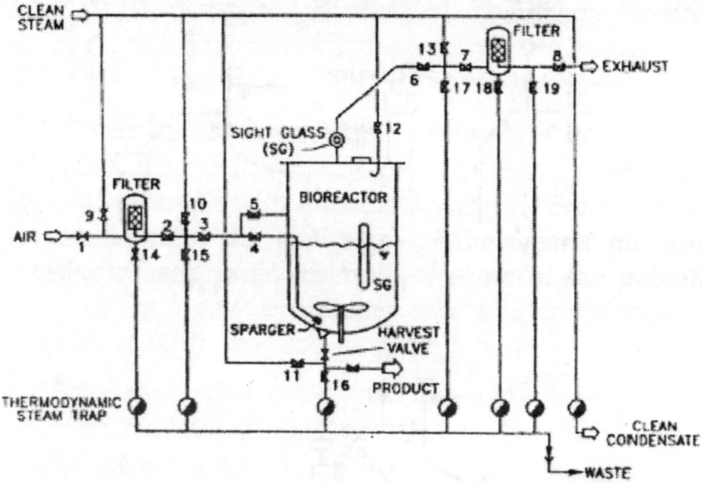


*Figure 6. Steam barrier at the harvest valve on a fermenter*

All contaminated or potentially contaminated condensate should drain to the biokill system. Closed systems should be sampled such that exposed surfaces are not contaminated, and no aerosols are released. This requirement must be met at the BL2-LS and higher. Needle-andsyringe sampling through rubber septa does not meet containment requirements.

Moreover, hypodermic needles have been frequently implicated in accidental inoculation of handling personnel. Contained sampling that releases no viable aerosols is detailed in Fig. 7. In contrast, an uncontained sampling system is depicted in Fig.8. The contained sampling principles detailed in Fig.7 can be incorporated in automated sampling devices one of which, available from Bioengineering AG, is shown in Fig.9. The sample container should be opened within a suitable biosafety cabinet.

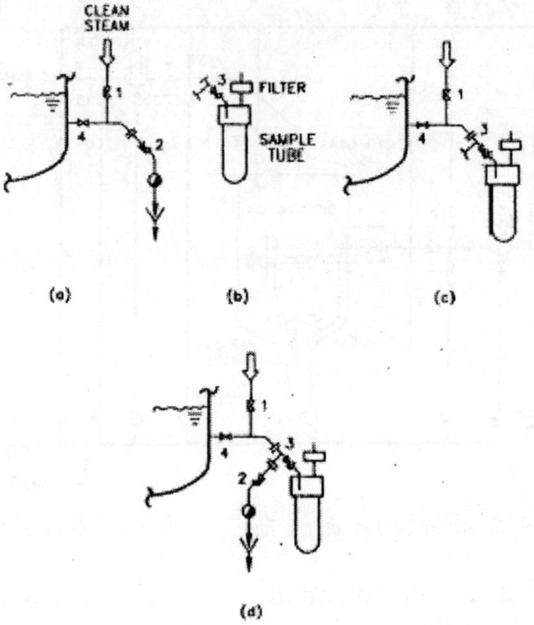


*Figure 7. Sterile, contained, aerosol-free sampling*

The acid and alkali reservoirs in BL3 -LS fermentation plants should be stainless steel pressure vessels that are hardpiped to the fermenter. For lesser level containment, glass reservoirs connected using silicone rubber or other similar tubing and peristaltic pumps are acceptable for pH control. Hardpiping, as opposed to using rapid connection couplings, is preferred for minimizing aerosol generation when high-level containment is necessary.

Culture transfer between fermenters should be through permanent hardpiped transfer lines in BL3 -LS facilities. Pertinent transfer practices have been described by Chisti. A transfer system meeting the BL2-LS criteria is shown in Fig.10. Although the needle-and-diaphragm type of connections are still frequently used, particularly during inoculation of fermenters, latest designs of safety connection devices (Fig.11) have virtually eliminated the need for needle-type connectors.

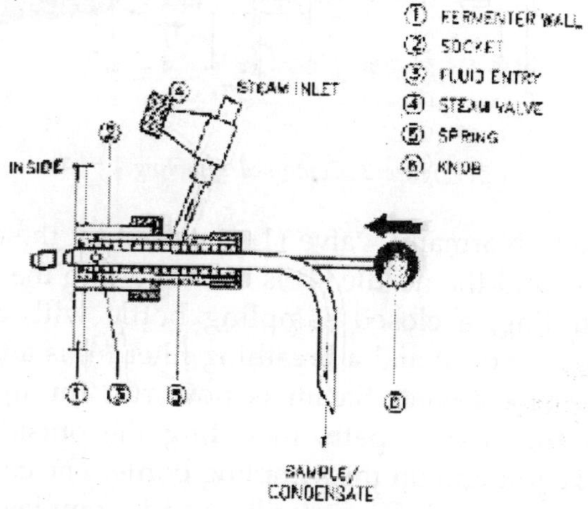


*Figure 8. Assembly for uncontained sampling*

Fully contained processing within closed equipment is feasible and has indeed been implemented, for example, in production of recombinant human interferon using E. coli K-12. Facilities manufacturing vaccines such as mumps, measles, rubella, varicella, or hepatitis need particular attention to containment.

Fermentater design practices noted by Chisti are generally sufficient for GLSP and BL1-LS operations, and can be easily extended to BL3-LS level (e.g., use of double mechanical seals; two exhaust filters in series). More general design issues relating to equipment for submerged culture and solid-state culture have been addressed elsewhere.

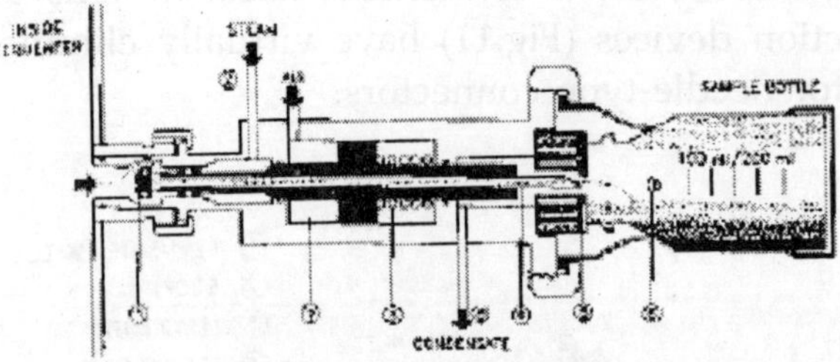


*Figure 9. Contained sampling.*

In figure 9, Normally, valve (1) is closed by the action of spring (3) and the needle (4) is retracted into the housing. For sampling, a closed sampling bottle with a rubber diaphragm cap (5) and a breathing filter (6) is attached to the sampling device. Steam is now run through (2) to sterilize the sample path, including the outside of the rubber diaphragm on the sampling bottle. The condensate is withdrawn at (8). Once sterilization is complete and the assembly has cooled, valve (1) is opened by pneumatic action of compressed air supplied to chamber (7).

The spring is compressed, the needle moves through the rubber diaphragm into the bottle, and sample flows in. Releasing the air pressure in chamber (7) closes the valve by the action of spring (3). The assembly must be re-sterilized and cooled before the sample bottle is removed. The entire sampling operation, including installation and removal of bottles can be automated, and sampling can be done at pre-programmed intervals.

### Downstream processing

During processing, the viable agent should be removed or inactivated as early as feasible in the process sequence. This usually means inactivation upon completion of fermentation, prior to downstream processing. Inactivation of E. coli K-12 using sulfuric acid in commercial processing has been noted. Alternatively, cells may be removed by microfiltration or ultrafiltration; however, operations such as centrifugation and macroporous filtration (e.g., filter presses) do not remove all viable particles.

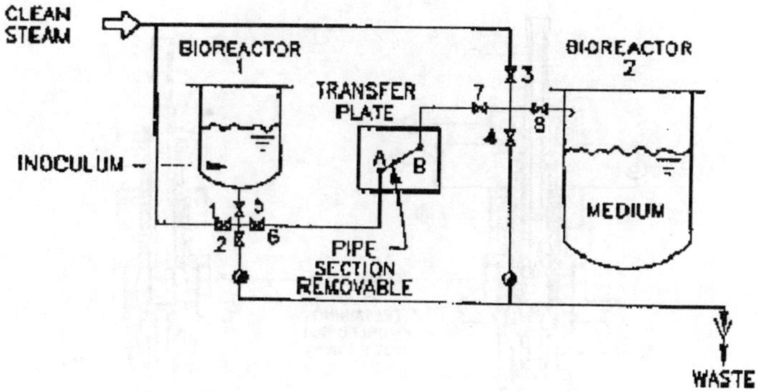


*Figure 10. Transfer System Meeting the BL2-LS standards*

Once the viable agent has been deactivated or removed, and the bioproduct poses no special risk, subsequent processing may proceed in open systems with due regard to the GMP-dictated requirements for protecting the product. When inactivation or removal are not feasible (e.g., live vaccines), downstream processing must be contained.

Containment of some downstream processing machinery - particularly centrifuges and cell-disruption devices - has been discussed by Deans and Stewart, and other commercially relevant mechanical cell-disruption equipment has been described. In addition to the viable agent, the physiological and toxicity profiles of products and byproducts constrain the choice of process equipment. Certain process schemes may be ruled out by the extent of containment needs and the amount and types of waste streams that would need to handled. In one case, a difficult-to-contain rotary drum filter that also generated difficult-todispose filter aid-mixed solids was replaced with fully contained ultrafiltration for recovery of cefoxitin.

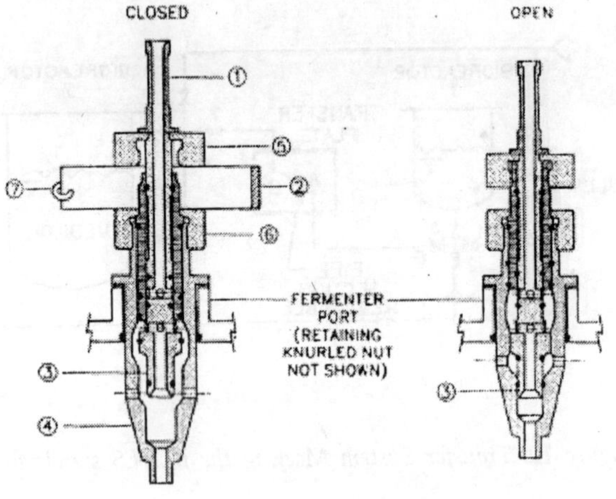


*Figure 11. A safety connection valve for coupling inoculum, acid, alkali, and antifoam reservoirs to the fermenter.*

In figure 11, the nipple (1) on the closed valve assembly is connected to transfer rubbing attached to an empty reservoir equipped with a breathing filter. The entire set-up (safety connection valve, transfer tube, and reservoir with filter) is autoclaved and cooled. The reservoir is filled with culture fluid inside a biological safety cabinet.

The valve body (4) is now installed in a port on top of the fermenter. The fermenter is sterilized after installation of all reservoirs (e.g., inoculum bottle, acid and alkali containers), and cooled. For inoculation, the protective clamp (2) is removed and the valve (3) is pushed in to allow pumping of the fluid into the fermenter. The system shown is satisfactory for GLSP processing, but not for higher-level containment.

Sometimes, primary containment within the process equipment may not be feasible, or contaminated machinery may have to be dismantled - for example, during harvesting of solids from tubular bowl batch centrifuges - and processing within enclosures would be necessary for primary containment. Isolation of aerosol-generating process equipment in HEPA-filter-exhausted, negativepressure enclosures is a suitable means of containment.

The containment room doors should be interlocked with the equipment so that the doors can be opened only when the equipment has stopped running, and sufficient time has elapsed for several space volumes of air to be exhausted from the contained area. Ideally, the equipment itself should be designed for primary containment, and isolation within enclosures should be an added safety measure to contain accidental release from high-pressure devices such as centrifuges and cell disruptors.

Performance of the primary and the secondary containment should be validated and monitored regularly. Multiproduct facilities may require product-dedicated process equipment to eliminate the likelihood of cross-contamination. This. is especially so for equipment that cannot be reasonably freed of all traces of a product. For example, chromatography media and membrane filters may have to be dedicated to specific products.

The use status - clean, in use, dirty, washed, sterile, etc. - of equipment should be clearly identified at all times. Downstream purification and formulation areas should process only one product at a time. A label control program must be in place. Handling of a concentrated, bioactive product can be the most dangerous part of bioprocessing. Fine powders are easily aerosolized; hence, handling of freezedried cultures or toxins can be particularly hazardous.

Because of their aerosol-generating potential, spray-dryers and freeze-dryers require special attention to containment. Freeze-drying of biohazardous substances has been discussed by Adams. As a general rule, automation and fully enclosed mechanized operation can significantly enhance containment while reducing the need for human-process contact.

*Other systems*

Heat exchange devices are frequently encountered in bioprocessing plants. Corroded heat exchangers (vessel jackets, plate heat exchangers, condensers, shell-and-tube exchangers, etc.) or those with leaking gaskets can contaminate the cooling fluid with the viable agent. Heat exchange equipment should be selected with regard to containment requirements, and cooling water may have to be monitored for contamination.

Clean-in-place (CIP) systems are another common feature of bioprocessing facilities. Automated CIP systems reduce exposure of personnel to hazardous material and assure consistent cleaning. The CIP system design has been treated by Chisti and Moo-Young. For contained facilities, fully closed CIP systems should be used. Attention should be given to cleanability and sterilizability of the CIP system itself, and generation of aerosols should be prevented. In view of the containment and cross-contamination considerations, certain process areas may require dedicated CIP systems.

***Personnel Protective Equipment***

Certain process operations are difficult to contain, and even the best designed primary containment can fail. Therefore, use of personnel protective clothing appropriate to risk is essential. Lab coats over street clothing are satisfactory for BL2-LS and lower-rated containment areas.

Correct gowning room practice is essential to preventing spread of contamination. Protective clothing - gowns, shoe coverings, head covering, face masks, and gloves - should be removed in the proper fashion before leaving the work area.

BL3-LS operations, for example, in hepatitis vaccine production, require that personnel remove all street garments down to underwear and don 'bunny suits', face masks, and gloves, prior to entering the work area. Protective clothing that completely isolates an individual from the environment is sometimes use. Depending on the characteristics of the product, such protective suites may be required even for BL2-LS or lower-rated processes.

*Figure 12. Universal biohazard sign.*

This type of isolation is provided by positive-pressure suits with battery-operated, forced air supply drawn from the surrounding environment through HEPA filters. Head-and-shoulder half suits with HEPA filtered air supply can also be used over other disposable protective clothing. The suits are equipped with low-battery alarms. Use of showers is necessary prior to leaving the degowning area of a BL3-LS biohazard containment zone.

Lower level containment areas should be equipped with hand washing facilities near the exit from the area. Sinks with automatic or hand- or foot-operated faucets (taps) are used. Suitable germicidal soap should be provided. Awareness of the biohazard is important to risk reduction. Consequently, BL2-LS and higher containment areas must display the universal biohazard sign (Fig. 12) on the entrance to the contained area. Additional information should include the containment level, the specific biohazard agent, any special entry requirements, and emergency contact details of responsible personnel.

#### *Personnel Training*

Personnel training is an essential part of safe bioprocessing. Even the best designed facilities, equipment, and practices will fail to provide the intended protection if the operators do not have the knowledge, the training, and the right attitude to personal safety, that of colleagues, the product, and the community.

Training should be provided in specific processing methods, operation of equipment, use of personnel protective equipment, gowning practices, aseptic and good microbiological technique, containment and biosafety measures consistent with the hazard, emergency procedures, and authorized practices.

Written operational protocols should identify the specific, actual or potential, hazards. Established practices should be strictly followed. Personnel should be supervised to assure consistent use of prescribed practices. Training in Good Manufacturing Practices is also required, and should be a continual process.

#### *Medical Surveillance*

Routine medical surveillance appropriate to risk is recommended. For example, electrocardiograms monitoring of individuals working with cardiotoxic substances, and seroconversion of individuals handling antigens. Although medical surveillance by itself does not protect against exposure, surveillance is useful in early detection and treatment. Surveillance also helps in identifying procedural or mechanical lapses.

Medical surveillance is especially necessary when pathogenic, potentially pathogenic, or new micro-organisms are being investigated, or when the nature of the hazard is unknown. In addition to the viable agent, the bioactivity of the product or any contaminants may pose a health hazard. Thus, for example, a nonpathogenic recombinant species may pose no risk beyond that associated with the corresponding wild strain, but the product of the inserted foreign gene may be highly bioactive, allergenic, or otherwise toxic.

When suitable vaccines are available, the process personnel as well as those providing support services but not directly working in the process areas (e.g., management, administration), should be immunized. A 'wait time' is necessary for development of immunity. Process workers should show resistance to the infective agent before being allowed into the work area. Furthermore, it should be recognized that vaccination may not guarantee protection against a high dose of the etiologic agent.

The workplace should be receptive to reporting of illnesses and accidents. Even apparently minor incidents - for example, being scratched while cleaning a process vessel - should be reported and recorded. Personnel who are ill, and those with open sores and cuts, should not be allowed into critical work areas. Immunocompromised individuals, those with diseases such as cancer and diabetes, those undergoing antimicrobial, steroid or immunosuppressive therapy are especially at risk. Attention to protection of peripheral support staff, for example the maintenance and cleaning personnel, is especially important because they may not have the knowledge or training for the potential risks.

***Biowaste***

All contaminated liquid, solid, and gaseous waste from a facility must be decontaminated prior to release. Solid waste is generally autoclaved or incinerated. Gases are filter sterilized and/or incinerated. Liquid effluent is collected into a containment sump and treated through the biokill system. All decontamination procedures should be validated, and the treated material should be examined for sufficiency of kill before being released.

The regulations relating to disposal and shipping of biohazardous wastes in the United States have been discussed by Tumberg. Effluent from the containment sump may be chemically disinfected or heat sterilized. Sterilization may be batchwise or continuous using direct steam injection or indirect heating. Good mixing of chemical additives and uniform sterilization temperature must be achieved for defined periods.

Waste decontamination areas are generally held at negative pressure which is mandatory when decontaminating BL3-LS effluent. There should be provisions for preventing accidental release of untreated material to sewer. For example, a locked effluent drain valve may be employed; the valve is opened only when a treated batch is released. Use of chemical disinfection prior to thermal sterilization is recommended to reduce the hazard in case of inadvertent effluent release.

The decontamination process may be automated. The sump and the sterilization tanks should be vented only through HEPA filters that may have to be heated to prevent condensation. The filtered exhaust from the waste decontamination tank may have to be treated for odor control if odor is a nuisance. Odor may be controlled by incineration, scrubbing, or absorption. Miller and Bergmann have discussed treatment of BL2-LS effluent.

In addition, the hygienic design practices noted by Chisti for fermentation plant apply also to biokill machinery. After decontamination, the biohazardous waste is generally disposed of using the same practices that apply to other nonbiological waste. Sometimes special treatment is necessary to destroy nonviable bioactive or otherwise environmentally harmful substances.

# Bibliography


B. Ekberg, K. Mosbach, "Molecular imprinting: a technique for producing separation materials", *Trends Biotechnol.* 7, 1989.

J. Silva, G. Weber, *Pressure stability of proteins,* Annu. Rev. Phys. Chem. 44, 1993.

M. Wilchek, T. Miron, J. Kohn, "Affinity chromatography", *Methods Enzymol*, 1984.

M.A. Vijayalakshmi, *Pseudobiospecific ligand affinity chromatography*, 7, 1989.

M.L. Yarmush, A.M. Weiss, K.P. Antonsen, D.J. Odde, D.M. Yarmush, "Immunoaffinity purification: basic principles and operational considerations", *Biotechnol.* Adv. 10, 1992.

P. Cuatrecasas, M. Wilchek, "Single-step purification of avidin from egg white by affinity chromatography on biocytin-Sepharose columns", *Biochem. Biophys.* Res. Commun. 33, 1968.

P. Cuatrecasas, M. Wilchek, C.B. Anfinsen, *Selective enzyme purification by affinity chromatography*, Proc. Natl. Acad. Sci. USA 61, 1968.

# Index